AF357334

Information Retrieval and Social Media Mining

Information Retrieval and Social Media Mining

Editor

María N. Moreno García

MDPI • Basel • Beijing • Wuhan • Barcelona • Belgrade • Manchester • Tokyo • Cluj • Tianjin

Editor
María N. Moreno García
University of Salamanca
Spain

Editorial Office
MDPI
St. Alban-Anlage 66
4052 Basel, Switzerland

This is a reprint of articles from the Special Issue published online in the open access journal *Information* (ISSN 2078-2489) (available at: https://www.mdpi.com/journal/information/special_issues/information_retrieval_social_media_mining).

For citation purposes, cite each article independently as indicated on the article page online and as indicated below:

LastName, A.A.; LastName, B.B.; LastName, C.C. Article Title. *Journal Name* **Year**, *Volume Number*, Page Range.

ISBN 978-3-0365-0246-5 (Hbk)
ISBN 978-3-0365-0247-2 (PDF)

Contents

About the Editor

María N. Moreno García is currently a Full Professor at the University of Salamanca, Spain, and head of the Data Mining Research Group (mida.usal.es). She has been the coordinator of the PhD program in Computer Engineering at the same University from 2013 to 2020. She has been a research scholar at the Intelligent System Lab of the University of Bristol, UK, and at the College of Computing and Digital Media of the DePaul University in Chicago, USA. She has organized six editions of the special session "Web mining and recommender systems" and three editions of the special session "Web and Social Media Mining" at the PAAMS conference and satellite events. She is a member of the editorial board of the journals *"Information"* and *"International Journal of Web Engineering and Technology"* and is a reviewer for journals indexed in relevant positions in the WoS Journal Citation reports. She is also the author of numerous articles published in recognized journals and conferences. Her research interests are in the areas of Data Science, Machine Learning, and their application in several domains, especially Decision Support in Medicine, Social Media, and Recommender Systems.

Preface to "Information Retrieval and Social Media Mining"

Many of today's businesses are taking advantage of advances in information retrieval and social media mining methods to increase their profits. These techniques allow them to personalize the products or services they offer their customers as well as to extract information from social networks to know user behavior, opinions, and sentiments, which can be exploited for multiple purposes.

This book aims to provide an insight into the progress made in the field of information retrieval and social media mining by presenting new contributions representative of the most recent research directions. They are focused on three highly topical areas: recommender systems, social media analysis, and sentiment analysis.

Since the first recommender systems appeared in the 1990s, research in this area has become increasingly interesting. Many methods have been proposed to provide users with personalized recommendations for products or services, although collaborative filtering (CF) is the most widespread approach. These techniques can be used alone or combined with other methods in hybrid approaches to tackle some problems that are specific to CF. A huge amount of current work is addressing the improvement of user recommendations in different ways. These range from the development of context-aware recommender systems or the evaluation of different aspects of the items to be recommended to the application of deep learning techniques, among others. Recently, the exploitation of social information is receiving special attention since social networks contain valuable data relating to user behavior, relations, interests, and preferences that can contribute to improving these systems. This book includes some proposals related to the mentioned topical issues.

Social networks have become a new source of virtually unlimited information that can be exploited through data analysis techniques in many domains, in addition to recommender systems. Every day their users generate, consume, and share through these media information about preferences, tastes, opinions, activities, location, relationships with other users, etc. The structure of these networks, their dynamics, the behavior of their users, the flow of information, etc. can be analyzed for diverse purposes. Some of them are the creation of user profiles, study of social influence, detection of implicit communities and analysis of their evolution, study of information diffusion, etc., which are subjects of unquestionable interest in many fields. In this task, social media mining plays a key role as the process of representing, analyzing, and extracting patterns from social media data. Some articles in the book are dedicated to the application of these techniques on social network data in order to obtain benefits in different areas of application.

Sentiment analysis and opinion mining are other areas of current intensive research in the domains of information retrieval and social media mining and have a wide range of applications. Their objective is to extract subjective information, such as positive, negative, or neutral opinions, from user-generated content through natural language processing, computational linguistics, and text mining techniques. Recently, deep learning models such as convolutional neural networks (CNNs) or recurrent neural networks (RNNs) have been used to improve their results. These methods require the text to be previously cleaned and transformed into numerical vectors by means of a preprocessing process that encompasses different tasks. In the last step of this process, the most used techniques are term frequency–inverse document frequency (TF–IDF) and word embedding, although the last approach is gaining increasing interest since it provides vectors capturing the word context, unlike other methods. In this regard, the development of word embedding techniques based on deep learning is also the focus of recent work. The latest articles in this book include proposals related to these topics of interest.

More detailed information on all the articles in the book is provided in the first, entitled "Information Retrieval and Social Media Mining", which gives an overview of each of them.

María N. Moreno García
Editor

 *information*

Editorial

Information Retrieval and Social Media Mining

María N. Moreno-García

Department of Computer Science and Automation, University of Salamanca, 37008 Salamanca, Spain; mmg@usal.es

Received: 9 December 2020; Accepted: 10 December 2020; Published: 11 December 2020

The large amount of digital content available through web sites, social networks, streaming services, and other distribution media, allows more and more people to access virtually unlimited sources of information, products, and services. This enormous availability makes it very difficult for users to find what they are really interested in. Hence, the great current interest in developing personalized methods of information retrieval as well as reliable recommendation algorithms that help users to filter and discover what fits their preferences.

Social networks are a big source of data, from which valuable information can be extracted by means of datamining algorithms. Social media mining allows us to explore a wide range of aspects regarding users, communities, networks structures, information diffusion and so on, to be further exploited in multiple domains.

This Special Issue includes important contributions to the field of information retrieval and social media mining. Specifically, the articles published focus on three areas of research of great interest at the present: recommender systems, social media analysis, and sentiment analysis.

Collaborative Filtering (CF) is the approach most extensively used in recommender systems. It requires either explicit or implicit user ratings for items to be recommended. Then, recommendations provided to a user are based on the ratings of other users with similar preferences. Usually, each item is valued globally with a single rating; however, there are application domains in which different aspects of the items are rated. In these cases, multi-criteria recommendation models are required. Among them, one of the most recent and successful proposals is the utility-based multi-criteria recommendation approach, in which different utility functions can be used to model the value of an item from the perspective of a user. In this issue, an improvement of these models is presented in a proposal [1] that addresses user over-/under-expectations on items through penalty-enhanced models. These involve penalties in the range of $[-1, 1]$ for over-expectations and under-expectations that are added to the utility score and are learned in conjunction with expectations in the same optimization process used to generate the top-N recommendations by maximizing the normalized discounted cumulative gain.

Sometimes, collaborative filtering methods are combined with content-based approaches to solve some problems of the former and obtain more reliable recommendations. This combination is used in a cascade hybrid proposal for document recommendation presented in this issue [2]. A content-based method that makes use of document processing techniques and document metadata is applied first to provide an initial list of recommendations. It also uses a function that involves term frequency (*tf*) and inverse document frequency (*idf*) weights for document ranking. In a second step, collaborative filtering is used to re-rank the previous list.

Research on recommender systems also benefits from the intensive work currently being done in the field of deep-learning algorithms. Deep neural networks are being used to overcome some problems associated with matrix factorization methods since they are able to better represent complex relations between users and items. However, their use is justified if the complexity of the problem or the number of instances of the training set is high. This is the scenario of a paper in this Special Issue [3], in which a graph convolutional network (GCN) algorithm called PharmaSage is proposed for providing pharmacy product cross-selling recommendations based on product feature information and sales data. The model was trained with a huge amount of real pharmaceutical data including almost a million products with complex properties and approximately 100 million sales transactions. This information is represented in a graph where each node represents a unique pharmacy product which also contains a vector encoding its descriptive data. Cross-selling for each pair of products is represented by undirected weighted edges between nodes. The GCN algorithm learns product embeddings by convolutions on aggregate neighborhood vectors. Finally, cosine similarity is applied to the output vectors to obtain recommendation scores.

Recommender systems are also one of the areas in which social data can be exploited to improve the reliability of recommendations. The incorporation of social functionalities in the recommender platforms has allowed their use in this domain. In [4], the concepts of trust and homophily derived from social structure are used to deal with the neighborhood bias of some CF recommendation methods which limits the number of items that can be recommended. Trust is derived from the friendship connections and is used to determine the degree of influence between users. Homophily is inferred from structural equivalence. This is a property often used to identify implicit communities in social networks. This is a way to capture the homophily concept since users belonging to the same community usually share interests and preferences. The similarities between users based on trust and homophily are used to extend the neighborhood of the active user and thus increase the number of potentially recommendable items.

Social media analysis is the focus of two articles in the Special Issue. One of them [5] presents a method for detecting significant events in social networks that can positively or negatively affect users. The changes in the user's followership network are used for event detection and are the base of a further analysis of the network dynamics. It is considered that an event for a given user takes place if the user experiences a follow burst or an unfollow burst in a time interval. To detect bursts, new follow/unfollow events are modeled as independent time series. Then, a time function representing the difference between the actual new follows/unfollows and the expected value for a given time is computed. A Personal Important Event (PIE) happens when the value of the function is higher than a threshold. The work also analyzes the evolution of the networks of users' followers and how the bursts caused by PIEs impact on the evolution.

The other paper focused on social media analysis presents a study about different aspects regarding the interrelationship of social media usage and perceived individual social capital [6]. A systematic procedure was applied to identify 80 scientific publications, which were analyzed in order to assess the measurement techniques used for evaluating social capital. Two operational techniques were identified. Additionally, the individual measurements items were explored to analyze future replication possibilities, resulting in no possibility of replication in an appreciable percentage of items. In the work, some consistencies and/or heterogeneity were detected in terms of operationalization, which can be useful for future studies.

In the research domains of information retrieval and social media mining, the application of language processing approaches to analyze sentiments is gaining increasing interest. In this context, the development of word embedding techniques based on deep learning have played an important role. In fact, word embedding is involved in a contribution to this issue [7], where sentiment analysis was performed for mining and summarizing opinions taking into account the context. The proposal, focused on news opinions, allows determining the relevance based not only on the text of the opinions, but also on the content of the news and its context. Topic detection from the opinion texts was performed by applying a hierarchical agglomerative clustering algorithm and using two different techniques to compute text similarity, with word embedding resulting as the best. The next steps are classifying the sentences according to the sentiment polarity and mapping topics and sentences. Finally, summary construction was provided after topic contextualization and sentence ranking were applied to news content. The topic was obtained by measuring the semantic similarity between the vocabulary associated with the topic and the news content.

We end this editorial by discussing another work that also addresses sentiment analysis [8]. In this case, the targets were questionnaire responses in telemonitoring programs to assist telemedicine patients. The aim was to monitor the adherence of patients to these programs from the sentiment polarity of their responses. The work presents the complete architecture of the system and also includes the collection and management of questionnaires. In addition, a new approach is introduced in the sentiment analysis that allows the monitoring of changes in patient's opinion across time through the repeated administration of a questionnaire. This is achieved by obtaining the polarity as a numerical value and modelling its sequence as a time series.

Funding: This work has been performed within the framework of a project funded by the Junta de Castilla y León, Spain, grant number SA064G19.

Conflicts of Interest: The author declares no conflict of interest. The funders had no role in the writing of the manuscript.

References

1. Zheng, Y. Penalty-Enhanced Utility-Based Multi-Criteria Recommendations. *Information* **2020**, *11*, 551. [CrossRef]
2. Borovič, M.; Ferme, M.; Brezovnik, J.; Majninger, S.; Kac, K.; Ojsteršek, M. Document Recommendations and Feedback Collection Analysis within the Slovenian Open-Access Infrastructure. *Information* **2020**, *11*, 497. [CrossRef]
3. Hell, F.; Taha, Y.; Hinz, G.; Heibei, S.; Müller, H.; Knoll, A. Graph Convolutional Neural Network for a Pharmacy Cross-Selling Recommender System. *Information* **2020**, *11*, 525. [CrossRef]
4. Sánchez-Moreno, D.; López, V.F.; Muñoz, M.D.; Sánchez, A.L.; Moreno, M.N. Exploiting the user social context to address neighborhood bias in collaborative filtering music recommender systems. *Information* **2020**, *11*, 439. [CrossRef]
5. Tang, T.; Hu, G. Detecting and Tracking Significant Events for Individuals on Twitter by Monitoring the Evolution of Twitter Followership Networks. *Information* **2020**, *11*, 450. [CrossRef]
6. Poecze, F.; Strauss, C. Social Capital on Social Media—Concepts, Measurement Techniques and Trends in Operationalization. *Information* **2020**, *11*, 515. [CrossRef]
7. Ramón-Hernández, A.; Simón-Cuevas, A.; García, M.M.; Arco, L.; Serrano-Guerrero, J. Towards Context-Aware Opinion Summarization for Monitoring Social Impact of News. *Information* **2020**, *11*, 535. [CrossRef]
8. Zucco, C.; Paglia, C.; Graziano, S.; Bella, S.; Cannataro, M. Sentiment Analysis and Text Mining of Questionnaires to Support Telemonitoring Programs. *Information* **2020**, *11*, 550. [CrossRef]

Publisher's Note: MDPI stays neutral with regard to jurisdictional claims in published maps and institutional affiliations.

 *information*

Article

Penalty-Enhanced Utility-Based Multi-Criteria Recommendations

Yong Zheng

Department of Information Technology and Management, College of Computing Illinois Institute of Technology, Chicago, IL 60616, USA; yzheng66@iit.edu

Received: 18 October 2020; Accepted: 20 November 2020; Published: 26 November 2020

Abstract: Recommender systems have been successfully applied to assist decision making in multiple domains and applications. Multi-criteria recommender systems try to take the user preferences on multiple criteria into consideration, in order to further improve the quality of the recommendations. Most recently, the utility-based multi-criteria recommendation approach has been proposed as an effective and promising solution. However, the issue of over-/under-expectations was ignored in the approach, which may bring risks to the recommendation model. In this paper, we propose a penalty-enhanced model to alleviate this issue. Our experimental results based on multiple real-world data sets can demonstrate the effectiveness of the proposed solutions. In addition, the outcomes of the proposed solution can also help explain the characteristics of the applications by observing the treatment on the issue of over-/under-expectations.

Keywords: recommender systems; utility; multi-criteria; penalty; over-expectation; under-expectation

---

1. Introduction

Information retrieval and recommender systems are two solutions to alleviate the problem of information overload [1], while recommender systems can deliver personalized recommendations to the end users without users' explicit queries. Recommender systems are usually built by learning from different types of the user preferences, such as explicit ratings or implicit feedbacks [2,3]. In the past decades, different types of the recommender systems have been proposed and developed. Multi-criteria recommender systems (MCRSs) [4] is one of these recommender systems which take the user preferences on different aspects of the items into account to improve the quality of the recommendations.

MCRSs have been implemented and served in real-world applications, such as hotel bookings at TripAdvisor.com, movie reviews at Yahoo!Movie, restaurant feedbacks at OpenTable.com. An example of the OpenTable.com can be shown by Figure 1. The system allows users to reserve tables at a restaurant and leave ratings on their dinning experiences. To review user experiences on a restaurant, we are able to observe the overall rating and multiple ratings on different aspects of the restaurant in Figure 1b, such as food, service, ambiance and noise level. It is because the system collects each user's overall rating and multi-criteria ratings as shown by Figure 1a. Afterwards, MCRSs can be built by taking advantage of these multi-criteria ratings in order to deliver more effective restaurant recommendations.

An example of data in MCRSs can be shown by Table 1. The rating refers to the users' overall rating on the items. We also have users' ratings on multiple criteria, such as food, service and value.

The research problem in MCRS is straightforward. Take the task of rating predictions for example; MCRSs predict an overall rating for a user and an item by taking advantage of the user's multi-criteria ratings on the item. In Table 1, MCRSs try to predict U_3's overall rating on T_1 as shown in the table above, while we do not know U_3's multi-criteria ratings on T_1. Usually, we need to estimate a user's multi-criteria ratings on an item, and then aggregate these ratings to finally predict the overall rating. The predicted overall rating can be used as a ranking score to sort and produce the list of recommendations delivered to the user.

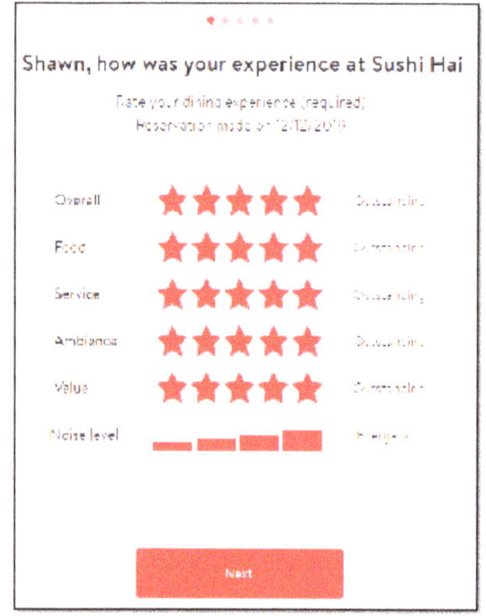

(**a**) Page of rating entry

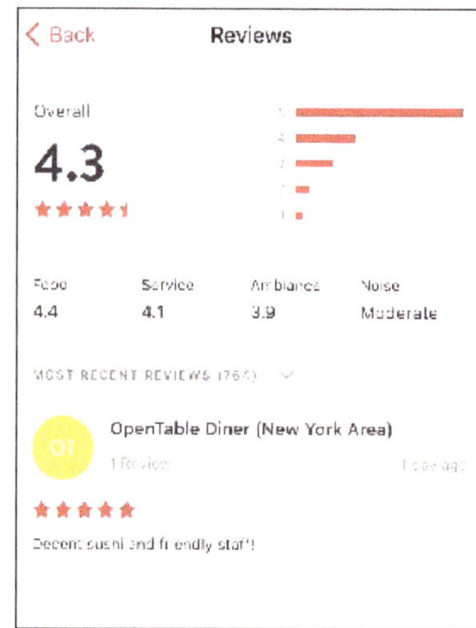

(**b**) Page of restaurant information

Figure 1. Example of user preferences on multiple criteria.

Table 1. Example of Rating Matrix from OpenTable.

User	Item	Rating	Food	Service	Value
U_1	T_3	4	4	3	4
U_2	T_2	3	3	3	3
U_3	T_1	?	?	?	?

Most recently, a utility-based multi-criteria recommendation approach [5] was proposed and it was demonstrated as one of the most effective methods. In this approach, we assume that there are user expectations on the items which can be represented by a list of ratings in multiple criteria. Given an item, we can also estimate a user's ratings on the different aspects of the items. In this case, the similarity between the user expectations and the multi-criteria ratings on the items can be considered as the utility of the item from the perspective of the user. A user may like the items more, if the similarity between user expectation and the user's multi-criteria ratings on these items is higher. The similarity score therefore can be used to rank the items to produce the top-N recommendations. We proposed to learn these user expectations by a learning-to-rank [6,7] method, and the experimental results were effective and promising.

However, there is a drawback in this approach. Namely, there is an issue of over-/under-expectations, while the current utility or similar function is not able to capture it. The issue refers to the situation that a user's rating on an item may lead to over-/under-expectations in comparison with the user's expectations on the items. Finally, It could result in false positives in the recommendation list and false negatives in the recommendation candidates. Take Table 2 for example, the first three rows refer to user u's rating vectors on three items, while the last row refers to u's expectations to select a restaurant to dine in. It is clear that u's ratings on T_1 are under-expectations, while his or her ratings on T_2 are over-expectations. However, some of u's ratings on T_3 are under-expectations, while others are over-expectations. It results in the difficulty of deciding whether the user will like T_3. It could be more complicated when it comes to the recommendation methods in the proposed utility-based multi-criteria recommendation models. A filtering strategy [8] may be helpful to alleviate the issue, but we need to pre-define the filtering rules by using domain knowledge. The challenge, therefore, becomes how to figure out a general solution for the utility-based multi-criteria recommendation model without domain knowledge.

Table 2. Example of over-/under-expectation.

User	Item	Food	Service	Value	Ambiance
u	T_1	2	2	2	2
u	T_2	4	4	4	4
u	T_3	1	4	2	1
u's expectation	3	3	3	3

In this paper, we propose to learn and apply penalties for the situation of over-/under-expectations. The proposed solution is generally enough to be applied in any applications, and we do not need any domain knowledge to define the filtering rules. The experimental results based on multiple data sets can demonstrate the effectiveness of our proposed solutions.

The remainder of this paper is organized as follows. Section 2 positions the related work. Section 3 presents the utility-based multi-criteria recommendation model. Section 4 discusses our proposed solution to alleviate the issue of over-/under-expectations. Section 5 presents the experimental results, followed by the conclusions and future work in Section 6.

2. Related Work

In this section, we discuss the related work in multi-criteria recommender systems, as well as the utility-based recommendation models.

2.1. Multi-Criteria Recommendations

As mentioned before, we have both overall rating and multi-criteria ratings in the rating data. The task in MCRS is predicting the overall rating for a user on an item by taking advantage of the multi-criteria ratings. Usually, we need to estimate a user's multi-criteria ratings on an item, and then aggregate these ratings to finally predict the overall rating, as shown in Equation (1). We use R_0 to represent the overall rating, and $R_{1,2,\cdots,k}$ as the multi-criteria ratings, while the function f is denoted as the aggregation function.

$$R_0 = f(R_1, R_2, \cdots, R_k) \tag{1}$$

Several multi-criteria recommendation algorithms have been developed to take advantage of these multi-criteria ratings. One of these methods is the heuristic approach [4,9] which utilizes the multi-criteria ratings to better calculate user-user or item-item similarities in the collaborative filtering algorithms. Another one is the model-based approach [4,10,11] which constructs a predictive model to estimate a user's overall rating on one item from the observed multi-criteria ratings. The model-based methods are usually more effective than the heuristic approach, since they are machine learning based algorithms which can even alleviate sparsity issues in the rating data.

Adomavicius, et al.'s [4] linear aggregation is one of the most basic and popular model which is usually utilized as a baseline for the purpose of benchmark. In this approach, we need to predict a user's rating on each criterion independently by using any rating function in the traditional recommender systems. Afterwards, we can use a linear regression as the aggregation function to finally estimate the overall rating by taking advantage of these predicted multi-criteria ratings.

One drawback in the approach above is that it ignores the correlation among the different criteria. Take the restaurant recommendation in the OpenTable for example, a user may not give a high rating on the criterion "value", if the user does not like the "food" in this restaurant. Researchers try to build more effective models by taking the correlation of the criteria into considerations. The flexible mixture model [10] is one of these attempts. It is a mixture model-based collaborative filtering algorithm incorporating the discovered dependency structure, while multiple criteria can be put on the structure connected with a user and an item by using two latent variables. We made another attempt and proposed the approach of criteria chains [11], in which we predicted the multi-criteria ratings in a sequence. The predicted preference in one criterion could be considered as contexts to be used to predict the preference in the next criterion. In this way, we were able to consider the correlation among criteria in the chain.

2.2. Utility-Based Recommendation Models

According to the classification of recommender systems by Burke [12], there are five categories—collaborative models [13,14], content-based recommenders [15,16], methods which utilize demographic information [17], knowledge-based algorithms [18,19], and utility-based models [5,20,21]. The utility-based recommenders make suggestions based on a computation of the utility of each item for the user. Utility can be used to indicate how valuable an item is from the perspective of a user. The utility function may vary from data to data, and there are no unified function to be generalized to different domains or applications. Guttman used different transformation functions (e.g., linear, square or universal functions) for different types of the attributes (e.g., continuous or discrete) in the context of online shopping [20]. Li et al. [22] defined the utility of recommending a potential link in the social networks by a linear aggregation of its value, cost, and the linkage likelihood. Moreover, Zihayat et al. proposed to use the aggregation of article-driven (e.g., popularity, topic distributions) and user-driven measures (e.g., clickstream, dwell time) as the utility function for news recommendations [21]. The utility-based multi-criteria recommendation model [5] discussed in the next section is an example which designs the utility function to serve multi-criteria recommendations. Different optimization methods can be applied to find the optimal solution in the utility-based recommendation model. A multi-objective optimizer [23,24] could be useful, if there are multiple objectives involved in the recommendation model.

Our previous work [5] proposed and developed the utility-based multi-criteria recommendation models. However, we ignored the over-/under-expectation issue. In this paper, we propose the improved solutions which are built upon the previous model but they further alleviate the issue of the over-/under-expectations.

3. Preliminary: Utility-Based Multi-Criteria Recommendations

In this section, we introduce the existing utility-based multi-criteria recommendation model [5].

3.1. Utility-Based Model (UBM)

The major contribution of our previous work [5] is the design of the utility function for the multi-criteria recommender systems. More specifically, the utility of an item from the perspective of the user refers to how valuable the item is in view of a user. It was defined as the similarity between the vector of user expectations and the vector of user ratings in the multiple criteria (i.e., different aspects of the items).

Assume there are N criteria, we use $\vec{c_u}$ to represent the vector of user expectations for a user u, and $\vec{r_{u,i}}$ denotes the u's rating vector (i.e., multi-criteria ratings) on the item i, as shown in Equations (2) and (3). Note that the expectation vector tells a user's expectations on the favorite items aligned to the same criteria used in the vector $\vec{r_{u,i}}$. More specifically, $r^t_{u,i}$ ($t = 1, 2, \cdots, N$) refers to user u's rating on the item i in the tth criterion. Accordingly, c_u^t can tell user u's expectation on the items in terms of the tth criterion. They must be in the same rating scale for each criterion.

$$\vec{c_u} = < c_u^1, c_u^2, \cdots, c_u^N > \tag{2}$$

$$\vec{r_{u,i}} = < r_{u,i}^1, r_{u,i}^2, \cdots, r_{u,i}^N > \tag{3}$$

The value of the utility can be obtained by the similarity or distance measures between two vectors, as shown in Equation (4). The larger the utility is, the more the user may like this item. Note that distance measure will represent dissimilarities, since the similarity will be higher if the distance is smaller.

$$Utility(u, i) = similarity(\vec{c_u}, \vec{r_{u,i}}) \tag{4}$$

Theoretically, any similarity measures can be applied in Equation (4), such as Pearson correlation, cosine similarity, or distance measures (e.g., Manhattan distance, Euclidean distance, etc.) as dissimilarity measures. Our research deliver more insights about these similarity measures. First of all, Pearson correlation may not be a good choice since the values may not be reliable if the number of dimensions in the vectors is limited. In the area of MCRS, we usually have three or four multiple criteria, which raises the concerns in Pearson correlation. In addition, the angle-based measures, such as the cosine similarity, are not appropriate, since it may produce 100% similarity if two vectors are parallel but with different values. As a result, the distance measures can be utilized to represent the dissimilarity. Any distance measures can be applied. We tried both Manhattan distance and the Euclidean distance, and found that we could get better results by using Euclidean distance. Therefore, we only present the results based on the Euclidean distance in this paper. The distance values should be normalized to the unit scale, and then we use 1 minus the normalized distance value to represent the similarity between the two vectors.

Therefore, the workflow in the utility-based recommendation model can be summarized as follows. We use the data in Table 1 for example, and our task is to produce the top-N recommendations to user U_3.

First of all, we need to make predictions on the multi-criteria ratings in order to obtain the vector of user ratings on the items, i.e., $\vec{r_{u,i}}$. In other words, we need to predict how U_3 will rate all candidate items on the three criteria, {food, service, value} in Table 1. In our work, we apply a process of independent predictions. More specifically, to predict how how U_3 will rate an item on the criterion "service", we will apply a traditional recommendation algorithm on the rating matrix <user, item, service>. Accordingly, we apply the same algorithm on other rating matrix associated with the ratings on each criterion. We use biased matrix factorization (BiasedMF) [25] as the recommendation algorithm in this step, since it is usually considered as a standard baseline and effective algorithm in the traditional recommender systems.

The rating prediction function by BiasedMF [25] can be shown in Equation (5).

$$\hat{r}_{ui} = \mu + b_u + b_i + p_u^T q_i \tag{5}$$

μ refers to the global average rating, while b_u and b_i are the user bias and item bias respectively. p_u and q_i are the latent-factor vector which can represent u and i respectively. The MF will learn these parameters by minimizing sum of squared errors by using stochastic gradient descent as the optimizer. The L_2 norms are usually added into the loss function as the regularization terms in order to alleviate overfitting. The loss function is described in Equation (6), where λ is the regularization rate. r_{ui} and $\hat{r}_{ui}$ are the real rating and predicted rating for the entry u, i. The model will learn from each entry u, i in the training set T. We use $p*, q*, b*$ to represent the user latent-factor vectors, item latent-factor vectors and biases respectively which are the parameters to be learned in the process of optimizations.

$$\underset{p*,q*,b*}{Minimize} \sum_{(u,i) \in T} (r_{ui} - \hat{r}_{ui})^2 + \lambda(||p_u||^2 + ||q_i||^2 + b_u^2 + b_i^2) \tag{6}$$

Once we obtain the users' predicted multi-criteria ratings on the items, we randomly initialize the expectation vector for each user, and learn these vectors by using the optimization below.

3.2. Optimization

We can initialize user expectations for each user at the beginning. In this case, we are able to use Equation (4) to calculate the utility score which will be used to rank the items to produce the top-N recommendations. Our previous work [5] learns these user expectations by maximizing the normalized discounted cumulative gain (NDCG) [26] which is a metric used for listwise ranking in the well-known learning-to-rank methods. Assuming each user u has a "gain" g_{ui} from being recommended an item i, the average discounted cumulative gain (DCG) for a list of J items is defined in Equation (7).

$$DCG = \frac{1}{N} \sum_{u=1}^{N} \sum_{j=1}^{J} \frac{g_{uij}}{max(1, log_b j))} \tag{7}$$

where the logarithm base is a free parameter, typically between 2 and 10. A logarithm with base 2 is commonly used to ensure all positions are discounted. NDCG is the normalized version of DCG given by Equation (8), where DCG^* is the ideal DCG, i.e., the maximum possible DCG.

$$NDCG = \frac{DCG}{DCG^*} \tag{8}$$

In terms of the listwise ranking, LambdaRank [27] can be applied to optimize NDCG directly. In addition, genetic and evolution algorithms have also been demonstrated as effective solutions in the listwise ranking optimization [28]. They have been utilized as the optimizer in the area of recommender systems before [29,30]. Our previous work found particle swarm optimization (PSO) [31] to be an effective optimizer, and it is easy to be implemented.

The basic workflow in the PSO can be described by Algorithm 1. In PSO, we need to initialize multiple particles to search for the optimal solution, while we use the NDCG shown in Equation (8) as the fitness function. The position of each particle is the parameters we need to learn. In our case, the position here refers to the all of the user expectation vectors. At the initialize stage, we need to define the number of particles, the initial positions and velocity. The velocity can define how much each particle can move (i.e., change the positions at the beginning).

Algorithm 1: Workflow in PSO.

---

initialization;
while $t <= MaxIteration$ **do**
    for *each particle* **do**
        Calculate fitness value;
        if *fitness is better than pBest* **then**
            | update pBest and its position;
        end
        if *fitness is better than gBest* **then**
            | update gBest and its position;
        end
    end
    for *each particle* **do**
        | update particle velocity according to Equation (9);
        | update particle position according to Equation (10);
    end
    | $t = t + 1$;
end

Each particle will run the algorithms with the initialized positions (i.e., user expectations) and velocity. The velocity is a vector with the same size of the position vector. For each run, we calculate the fitness value, where it refers to the NDCG metric in our experiments. The learning process will save a cBest value (i.e., the best NDCG for each particle c in multiple runs) for each particle and a gBest value (i.e., the best NDCG by the whole group of the articles) for the whole group, as well as their corresponding positions. In each iteration, the process will update the velocity for each particle, as shown in Equation (9). We use $V_{ij,t}$ to denote the velocity of the jth bit in the position of the particle i in the tth learning iteration, $X_{ij,t}$ as the value of position in the jth bit in particle i in the tth iteration. P_{cBest} and P_{gBest} are the vector of positions associated with the individual best fitness (i.e., cBest) and the global fitness value (i.e., gBest). w_t, α_1, α_2, φ_1 and φ_2 are the arguments to be defined in advance. In this way, each particle can learn from itself and the best move by the whole group in each learning iteration.

$$V_{ij,t} = w_t \times V_{ij,t} + \alpha_1 \varphi_1 \times (P^j_{cBest} - X_{ij,t}) + \alpha_2 \varphi_2 \times (P^j_{gBest} - X_{ij,t}) \tag{9}$$

Finally, the position of each particle can be updated by Equation (10) and be used in the next learning iteration.

$$X_{ij,t+1} = X_{ij,t} + V_{ij,t} \tag{10}$$

4. Penalty-Enhanced Utility-Based Multi-Criteria Recommendation Model

In this section, we point out the issue of over-/under-expectation in the approach above, and discuss out solution which applies a penalty in the learning process.

4.1. Issue of Over-/Under-Expectations

To better explain the issue of over-/under-expectations, we use the example shown in Table 2. The first three rows present a user u's predicted rating vectors $\overrightarrow{r_{u,i}}$ on three items—T_1, T_2, T_3. The last row gives the user expectation vector $\overrightarrow{c_u}$.

For simplicity, we use the Manhattan distance to represent the dissimilarity between two vectors. In this case, the Manhattan distance is 4 which is the same for the items T_1 and T_2. Apparently, the ratings on the item T_2 are all above the user expectations, while the ratings on T_1 are all below the user expectations. Without solving the issue of over-/under-expectations, the items T_1 and T_2 will be considered equally in the item rankings. The situation could be more complicated. Take the item T_3 for example, the Manhattan distance will be 6 for T_3, but it falls in over-expectation in the criterion "Room", and under-expectation in other criteria. T_3 will be ranked ahead T_1 and T_2, but the end user may prefer T_2 rather than T_3. As a result, there could be false positives in the recommendation list and false negatives in the list of recommendation candidates.

We realized this issue, and proposed to use a filtering strategy to alleviate this issue [8]. More specifically, we can pre-define the rules for over-/under-expectations. For example, if the item falls in the situation of over-expectations, we may exclude this item from the list of candidate items to be recommended. However, it is difficult to pre-define these rules without domain knowledge, since we do not know whether the user will like an item if it falls in the case of over-expectation or under-expectation. In this paper, we seek solutions which are general and independent of domain knowledge.

4.2. Penalty-Enhanced Models (PEMs)

Our solution is simple and straightforward. We plan to learn a "penalty" for each situation. We define P_{over} and P_{under} as the penalty for the situation of over-expectation and under-expectations respectively. Everytime when we produce the utility score, we will add these penalties according to whether the actual situation is either over- or under-expected. The scale of P_{over} and P_{under} is $[-1, 1]$, since the utility score that was measured by similarity will fall in [0, 1]. We are going to learn P_{over} and P_{under} together with the user expectations in the learning-to-rank process.

Note that, we name it as "penalty", but actually the value could be positive or negative. It is a real penalty if the value is negative, since we will penalize the utility score. Otherwise, it is a bonus which will add values to the utility score—it implies that we still accept the item and it provides extra value in the situation of over- or under-expectations.

The remaining challenge is how to detect the situation of over- and under-expectations. We use a sign which can be computed by using $\sum_{t=1}^{N}(\vec{c_u^t} - \vec{r_{u,i}^t})$. The item is under-expected if the sign is positive. Otherwise, it is over-expected, if the sign is negative. We will not apply any penalties if the sign is zero.

A finer-grained approach is to learn these penalties for each user or each group of the users, since the penalties may vary from user to user. Learning the penalties for each user may suffer the sparsity problem In this paper, we use PEM+ to denote the approach that we learn P_{over} and P_{under} for each group of the users in our experiments, while we create the user groups by using the K-Means clustering [32] technique.

5. Experiments and Results

In this section, we present our data sets, evaluation strategies and the experimental results.

5.1. Data Sets and Evaluations

We use four real-world data set with multi-criteria ratings:

- TripAdvisor data: This data was crawled by Jannach, et al. [33]. The data was collected through a Web crawling process which collects users' ratings on hotels located in 14 global metropolitan destinations, such as London, New York, Singapore, etc. There are 22,130 ratings given by 1502 users and 14,300 hotels. Each user gave at least 10 ratings which are associated with multi-criteria ratings on seven criteria: value for the money, quality of rooms, convenience of the hotel location, cleanliness of the hotel, experience of check-in, overall quality of service and particular business services.

- Yahoo!Movie data: This data was obtained from YahooMovies by Jannach, et al. [33]. There are 62,739 ratings given by 2162 users on 3078 movies. Each user left at least 10 ratings which are associated with multi-criteria ratings on four criteria: direction, story, acting and visual effects.

- SpeedDating data: It was available on Kaggle (https://www.kaggle.com/annavictoria/speed-dating-experiment). There are 8378 ratings given by 392 users. It is a special data for reciprocal people-to-people recommendations, while the "items" to be recommended are the users too. Each user will rate his or her dating partner in six criteria: attractiveness, sincerity, intelligence, fun, ambition, and shared interests.

- ITMLearning data: It was collected for the educational project recommendations [34], while the authors used the filtering strategy to alleviate the over-/under-expectations. There are 3306 ratings given by 269 users on 70 items. Each rating entry is also associated with three criteria: app (how students like the application of the project), data (the ease of data preprocessing in the project) and ease (the overall ease of the project).

We compare the proposed PEM and PEM+ approaches with the following baseline approaches:

- The matrix factorization (MF) is the biased matrix factorization model [25] by using the rating matrix <User, Item, Ratings> only without considering multi-criteria ratings.

- The linear aggregation model (LAM) [4] is a standard aggregation-based multi-criteria recommendation method which predicts the multi-criteria ratings independently and uses a linear aggregation to estimate a user's overall rating on an item.

- The criteria chain model (CCM) [11] and flexible mixture model (FMM) [10] are two methods which take the correlation among criteria into consideration.
- The UBM model which is the original utility-based multi-criteria recommendation model without handling the over-/under-expectation issues.

We apply 5-fold cross validation on these data sets, and evaluate the performance of recommendations based on top-10 recommendations by using precision and NDCG. Furthermore, we use the particle swarm optimization (PSO) [35] as introduced previously. Particularly, we use OMOPSO [36] in the open-source library MOEA (http://moeaframework.org). OMOPSO was demonstrated as one of the top-performing PSO algorithms. MOEA is an open-source library for multi-objective learning, but it can also be used for single-objective learning, while we just setup NDCG as the only objective in the library. MOEA provides built-in optimal parameters for each learning algorithm, and we use these default parameters.

In addition to the PEM approach discussed in Section 4.2, we also examine PEM+ in which we put users into different clusters and learn the penalties for each cluster of the users. More specifically, we use the classical K-Means clustering on the user-item rating matrix. We tried different values for K (K = 2, 4, 6, 8, 10), and we found that the optimal value of K is 8, 6, 4, 4 for the TripAdvisor, Yahoo!Movie, SpeedDating and ITMLearning data respectively by using the the within-cluster sum of squared errors. We would like to examine whether PEM+ can offer further improvements, we just tried the small K values. The performance could be better if we try larger values, while we may also have more parameters to be learned. In PEM+, we will learn P_{over} and P_{under} for each cluster of users.

5.2. Results and Findings

First of all, we present the results based on precision and NDCG in Figure 2. Table 3 presents the NDCG results for the utility-based recommendation models, as well as the improvement by PEM and PEM+ in comparison with UBM. We performed two-paired t-test as the significant test at the 95% confidence level. We use * to represent significant results between proposed approach (i.e., PEM and PEM+) and the best performing baseline method, and ∘ to indicate significant results between PEM and PEM+. Significance results based on precision are depicted in Figure 2, while the results for NDCG are described in Table 3.

First of all, we compared the results among the baseline methods (i.e., MF, LAM, FMM, CCM and UBM). We observed that the UBM approach generally outperformed other baseline methods in terms of both precision and NDCG. UBM produced slightly better NDCG results than the NDCG by FMM in the TripAdvisor and Yahoo!Movie data.

By comparing the solutions proposed in this paper (i.e., PEM and PEM+) with the baseline methods, we observed that the PEM could offer improvements on both precision and NDCG on all the data sets, except the speed dating data. PEM+ was able to beat all baselines except the speed dating data too. We believe that the failure was caused by the characteristics of this data set, which will be discussed in the next paragraph. A further look at the comparison between PEM and PEM+ can tell that PEM+ beat PEM in NDCG for all data except the dating data. However, PEM+ failed to outperform PEM in precision for the Yahoo!Movie and ITMLearning data. Recall that we used the NDCG as the fitness function in PSO, while the results on precision may be out of controls. Another potential reason could be that we did not try larger K values in KMeans for PEM+.

As a summary, PEM and PEM+ could offer improvements over the utility-based recommendation model. The only exception was the SpeedDating data set. We did have multi-criteria ratings in this data set. However, it was a data set for people-to-people recommendations which fell in the category of reciprocal recommendations. The nature of this data was different from other multi-criteria rating data, which may have resulted in less improvements here. We observed that the NDCG was even decreased by using PEM. The underlying reasons may lie in the special characteristics of the reciprocal recommendations. In the context of speed dating, a successful recommendation will consider a "match" between two users. In our recommendation approach, we only considered the preferences from the perspective of the users who received the recommendations, but ignored whether the recommended people would like to date with the target user. It may result in a drop or less improvements. A reciprocal recommendation model which also considers the dating partners [37,38] may help improve the recommendation performance.

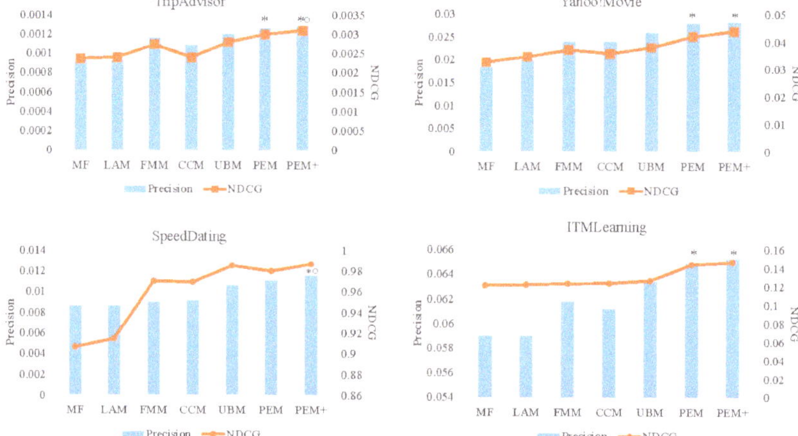

Figure 2. Experimental results.

Table 3. Results based on normalized discounted cumulative gain (NDCG).

	TripAdvisor	**Yahoo!Movie**	**SpeedDating**	**ITMLearning**
UBM	0.0028	0.038	0.9852	0.1264
PEM	0.003 (7.14%) *	0.042 (10.5%) *	0.98 (−0.5%)	0.1441 (14%) *
PEM+	0.0031 (10.7%) * ○	0.044 (15.8%) * ○	0.9866 (0.14%)	0.1466 (15.9%) * ○

Our previous research [8] proposed to use the filtering strategies to alleviate the issue of over-/under-expectations for the ITMLearning data. We chose the best filtering strategy and run the model. It achieved the NDCG result as 0.1311 which was lower than the results by using both PEM and PEM+. It is not surprising, since the filtering operation may mistakenly remove the items that a user may like. Our solution based on the penalties actually provided a soft and finer-grained solution to alleviate the issue of over-/under-expectations. These results demonstratde that our solution was much more effective than the filtering strategy, not to mention that the penalty-enhanced solution did not require any domain knowledge to define the rules for filtering.

Finally, we present the learned P_{over} and P_{under} by using the PEM approach, as shown by Table 4.

We observed that the penalties learned by our models varied from case to case. The "penalty" was positive for over-expectations and negative for under-expectations for the TripAdvisor, Yahoo!Movie and ITMLearning data sets. It tells that the users still liked the item if it was over-expected, and additionally a bonus was added to the predicted score which was used to rank the items. The penalty was negative in the case of under-expectation, so the predicted score was penalized accordingly. The pattern in the SpeedDating data was different from others—the penalty for over-expectation was negative, while it was positive for under-expectations. It implies that a user may not have accepted a recommended partner if some characteristics of the partner were over-expected. By contrast, the penalty for under-expectation was positive but close to zero, which implies that a partner was still acceptable even if the partner slightly missed the expectations in some characteristics.

These results are interesting and can also help us understand more characteristics about each data or domain.

Table 4. Learned penalties.

	P_{over}	P_{under}
TripAdvisor	0.124	-0.022
Yahoo!Movie	0.574	-0.985
SpeedDating	-0.29	0.02
ITMLearning	0.324	-0.165

6. Conclusions and Future Work

In this paper, we point out the issue of over-/under-expectations in the existing utility-based multi-criteria recommendation approach, and propose to learn penalties to alleviate this issue. Our experimental results based on four real-world data sets can demonstrate the effectiveness of the proposed solutions. Particularly, the penalty-enhanced approach works better than the filtering strategy, and it is general enough to be applied to any data sets.

However, there are still some limitations in the current work. We can consider more solutions for these issues as our future work. First of all, we define the case of over-/under-expectation for each rating entry by a user on an item, and apply the corresponding penalties. We can actually exploit a finer-grained method which will apply a penalty to each bit of the rating vector (i.e., case by case for the rating on each criterion). In this case, we have more penalties to be learned, but it may be able to further improve the models. In addition, we did not try larger K values for the KMeans clustering in the PEM+ method. Other K values may deliver better results. Using PSO as the optimizer may result in an efficiency issue for a large-scale data. We can use cloud service (such as Amazon Web Services) to learn the parameters. Or, we can seek other optimization methods in future. Finally, the penalties may be affected by other information, such as contexts [39,40] or trust information [41,42]. For example, the issue of over-/under-expectations may be serious in some contexts, but they can be ignored in other situations. Or, the issue can be ignored if the item was recommended by a trusted person. We will seek these alternative improvements in our future work.

Funding: This research received no external funding.

Conflicts of Interest: The author declares no conflict of interest.

Abbreviations

The following abbreviations are used in this manuscript:

CCM	Criteria Chain Model
DCG	Discounted Cumulative Gain
FMM	Flexible Mixture Model
LAM	Linear Aggregation Model
MCRS	Multi-Criteria Recommender Systems
MF	Matrix Factorization
MOEA	Multi-Objective Evolutionary Algorithms
NDCG	Normalized Discounted Cumulative Gain
PEM	Penalty-Enhanced Model
PSO	Particle Swarm Optimization
UBM	Utility-Based Model

References

1. Bawden, D.; Robinson, L. The dark side of information: Overload, anxiety and other paradoxes and pathologies. *J. Inf. Sci.* **2009**, *35*, 180–191. [CrossRef]
2. Alexandridis, G.; Siolas, G.; Stafylopatis, A. ParVecMF: A paragraph vector-based matrix factorization recommender system. *arXiv* **2017**, arXiv:1706.07513.
3. Alexandridis, G.; Tagaris, T.; Siolas, G.; Stafylopatis, A. From Free-text User Reviews to Product Recommendation using Paragraph Vectors and Matrix Factorization. In *Companion Proceedings of the 2019 World Wide Web Conference*; Association for Computing Machinery: New York, NY, USA, 2019; pp. 335–343.
4. Adomavicius, G.; Kwon, Y. New recommendation techniques for multicriteria rating systems. *IEEE Intell. Syst.* **2007**, *22*, 48–55. [CrossRef]
5. Zheng, Y. Utility-based multi-criteria recommender systems. In Proceedings of the 34th ACM/SIGAPP Symposium on Applied Computing, Limasso, Cyprus, 8–12 April 2019; pp. 2529–2531.
6. Liu, T.Y. *Learning to Rank for Information Retrieval*; Springer Science & Business Media: Berlin/Heidelberg, Germany, 2011.
7. Balakrishnan, S.; Chopra, S. Collaborative ranking. In Proceedings of the Fifth ACM International Conference on Web Search and Data Mining, Seattle, WA, USA, 8–12 February 2012; pp. 143–152.
8. Zheng, Y.; Ghane, N.; Sabouri, M. Personalized Educational Learning with Multi-Stakeholder Optimizations. In Proceedings of the Adjunct ACM Conference on User Modelling, Adaptation and Personalization, Larnaca, Cyprus, 9–12, June 2019.
9. Manouselis, N.; Costopoulou, C. Experimental analysis of design choices in multiattribute utility collaborative filtering. *Int. J. Pattern Recognit. Artif. Intell.* **2007**, *21*, 311–331. [CrossRef]
10. Sahoo, N.; Krishnan, R.; Duncan, G.; Callan, J. Research Note—The Halo Effect in Multicomponent Ratings and Its Implications for Recommender Systems: The Case of Yahoo! Movies. *Inf. Syst. Res.* **2012**, *23*, 231–246. [CrossRef]
11. Zheng, Y. Criteria Chains: A Novel Multi-Criteria Recommendation Approach. In Proceedings of the 22nd International Conference on Intelligent User Interfaces, Limassol, Cyprus, 13–16 March 2017; pp. 29–33.
12. Burke, R. Hybrid recommender systems: Survey and experiments. *User Model. User-Adapt. Interact.* **2002**, *12*, 331–370. [CrossRef]
13. Schafer, J.B.; Frankowski, D.; Herlocker, J.; Sen, S. Collaborative filtering recommender systems. In *The Adaptive Web*; Springer: Berlin/Heidelberg, Germany, 2007; pp. 291–324.
14. Ekstrand, M.D.; Riedl, J.T.; Konstan, J.A. Collaborative Filtering Recommender Systems. Available online: https://dl.acm.org/doi/10.1561/1100000009 (accessed on 23 November 2020).
15. Pazzani, M.J.; Billsus, D. Content-based recommendation systems. In *The Adaptive Web*; Springer: Berlin/Heidelberg, Germany, 2007; pp. 325–341.
16. Lops, P.; De Gemmis, M.; Semeraro, G. Content-based recommender systems: State of the art and trends. In *Recommender Systems Handbook*; Springer: Berlin/Heidelberg, Germany, 2011; pp. 73–105.
17. Zhao, W.X.; Li, S.; He, Y.; Wang, L.; Wen, J.R.; Li, X. Exploring demographic information in social media for product recommendation. *Knowl. Inf. Syst.* **2016**, *49*, 61–89. [CrossRef]
18. Burke, R. Knowledge-based recommender systems. *Encycl. Libr. Inf. Syst.* **2000**, *69*, 175–186.
19. Tarus, J.K.; Niu, Z.; Mustafa, G. Knowledge-based recommendation: A review of ontology-based recommender systems for e-learning. *Artif. Intell. Rev.* **2018**, *50*, 21–48. [CrossRef]
20. Guttman, R.H. Merchant Differentiation through Integrative Negotiation in Agent-Mediated Electronic Commerce. Ph.D. Thesis, Massachusetts Institute of Technology, Cambridge, MA, USA, 1998.
21. Zihayat, M.; Ayanso, A.; Zhao, X.; Davoudi, H.; An, A. A utility-based news recommendation system. *Decis. Support Syst.* **2019**, *117*, 14–27. [CrossRef]
22. Li, Z.; Fang, X.; Bai, X.; Sheng, O.R.L. Utility-based link recommendation for online social networks. *Manag. Sci.* **2017**, *63*, 1938–1952. [CrossRef]
23. Ribeiro, M.T.; Lacerda, A.; Veloso, A.; Ziviani, N. Pareto-efficient hybridization for multi-objective recommender systems. In Proceedings of the sixth ACM conference on Recommender systems, Dublin, Ireland, 9–13 September 2012.
24. Ribeiro, M.T.; Ziviani, N.; Moura, E.S.D.; Hata, I.; Lacerda, A.; Veloso, A. Multiobjective pareto-efficient approaches for recommender systems. *ACM Trans. Intell. Syst. Technol.* **2014**, *5*, 1–20.

25. Koren, Y.; Bell, R.; Volinsky, C. Matrix factorization techniques for recommender systems. *Computer* **2009**, *42*, 30–37. [CrossRef]

26. Valizadegan, H.; Jin, R.; Zhang, R.; Mao, J. Learning to Rank by Optimizing NDCG Measure. Available online: https://dl.acm.org/doi/10.5555/2984093.2984304 (accessed on 23 November 2020).

27. Donmez, P.; Svore, K.M.; Burges, C.J. On the local optimality of LambdaRank. In Proceedings of the 32nd International ACM SIGIR Conference on Research and Development in Information Retrieval, Boston, MA, USA, 19–23 July 2009; pp. 460–467.

28. Yeh, J.Y.; Lin, J.Y.; Ke, H.R.; Yang, W.P. Learning to rank for information retrieval using genetic programming. In Proceedings of the SIGIR 2007 Workshop on Learning to Rank for Information Retrieval (LR4IR 2007), Amsterdam, The Netherlands, 23–27 July 2007.

29. Ujjin, S.; Bentley, P.J. Particle swarm optimization recommender system. In Proceedings of the 2003 IEEE Swarm Intelligence Symposium, SIS'03 (Cat. No. 03EX706), Indianapolis, IN, USA, 26 April 2003; pp. 124–131.

30. Zheng, Y.; Burke, R.; Mobasher, B. Recommendation with differential context weighting. In Proceedings of the International Conference on User Modeling, Adaptation, and Personalization, Rome, Italy, 10–14 June 2013; Springer: Berlin/Heidelberg, Germany, 2013; pp. 152–164.

31. Poli, R.; Kennedy, J.; Blackwell, T. Particle swarm optimization. *Swarm Intell.* **2007**, *1*, 33–57. [CrossRef]

32. Kanungo, T.; Mount, D.M.; Netanyahu, N.S.; Piatko, C.D.; Silverman, R.; Wu, A.Y. An efficient k-means clustering algorithm: Analysis and implementation. *IEEE Trans. Pattern Anal. Mach. Intell.* **2002**, *24*, 881–892. [CrossRef]

33. Jannach, D.; Zanker, M.; Fuchs, M. Leveraging multi-criteria customer feedback for satisfaction analysis and improved recommendations. *Inf. Technol. Tour.* **2014**, *14*, 119–149. [CrossRef]

34. Zheng, Y. Personality-Aware Decision Making In Educational Learning. In Proceedings of the 23rd International Conference on Intelligent User Interfaces, Tokyo, Japan, 7–11 March 2018; p. 58.

35. Shi, Y.; Eberhart, R.C. Empirical study of particle swarm optimization. In Proceedings of the 1999 Congress on Evolutionary Computation-CEC99, Washington, DC, USA, 6–9 July 1999; Volume 3, pp. 1945–1950.

36. Sierra, M.R.; Coello, C.A.C. Improving PSO-based multi-objective optimization using crowding, mutation and ϵ-dominance. In Proceedings of the International Conference on Evolutionary Multi-Criterion Optimization, Guanajuato, Mexico, 9–11 March 2005; Springer: Berlin/Heidelberg, Germany, 2005; pp. 505–519.

37. Pizzato, L.; Rej, T.; Chung, T.; Koprinska, I.; Kay, J. RECON: A reciprocal recommender for online dating. In Proceedings of the Fourth ACM Conference on Recommender Systems, Barcelona, Spain, 26–30 September 2010; pp. 207–214.

38. Zheng, Y.; Dave, T.; Mishra, N., Kumar, H. Fairness In Reciprocal Recommendations: A Speed-Dating Study. In Proceedings of the Adjunct ACM Conference on User Modelling, Adaptation and Personalization, Singapore, 8–11 July 2018.

39. Adomavicius, G.; Mobasher, B.; Ricci, F.; Tuzhilin, A. Context-Aware Recommender Systems. *AI Mag.* **2011**, *32*, 67–80. [CrossRef]

40. Adomavicius, G.; Tuzhilin, A. Context-aware recommender systems. In *Recommender Systems Handbook*; Springer: Berlin/Heidelberg, Germany, 2011; pp. 217–253.

41. Agreste, S.; De Meo, P.; Ferrara, E.; Piccolo, S.; Provetti, A. Trust networks: Topology, dynamics, and measurements. *IEEE Internet Comput.* **2015**, *19*, 26–35. [CrossRef]

42. Lee, J.; Noh, G.; Oh, H.; Kim, C.k. Trustor clustering with an improved recommender system based on social relationships. *Inf. Syst.* **2018**, *77*, 118–128. [CrossRef]

Publisher's Note: MDPI stays neutral with regard to jurisdictional claims in published maps and institutional affiliations.

 *information*

Article

Document Recommendations and Feedback Collection Analysis within the Slovenian Open-Access Infrastructure

Mladen Borovič *, Marko Ferme, Janez Brezovnik, Sandi Majninger, Klemen Kac and Milan Ojsteršek

Faculty of Electrical Engineering and Computer Science, University of Maribor, 2000 Maribor, Slovenia; marko.ferme@um.si (M.F.); janez.brezovnik@um.si (J.B.); sandi.majninger@um.si (S.M.); klemen.kac@um.si (K.K.); milan.ojstersek@um.si (M.O.)
* Correspondence: mladen.borovic@um.si; Tel.: +386-2-220-74-60

Received: 24 September 2020; Accepted: 21 October 2020; Published: 23 October 2020

Abstract: This paper presents a hybrid document recommender system intended for use in digital libraries and institutional repositories that are part of the Slovenian Open Access Infrastructure. The recommender system provides recommendations of similar documents across different digital libraries and institutional repositories with the aim to connect researchers and improve collaboration efforts. The hybrid recommender system makes use of document processing techniques, document metadata, and the similarity ranking function BM25 to provide content-based recommendations as a primary method. It also uses collaborative-filtering methods as a secondary method in a cascade hybrid recommendation technique. We also provide a real-world data feedback collection analysis for our hybrid recommender system on an academic digital repository in order to be able to identify suitable time-frames for direct feedback collection during the year.

Keywords: hybrid recommender systems; feedback collection; digital libraries; information retrieval; real-world data; open-access

---

1. Introduction

Recommender systems are a part of everyday experience on the web, especially while using online stores and search engines. The main objective of these systems is to provide the user with relevant and interesting content. In digital repositories, the obvious task for a recommender system is to provide recommendations to relevant documents. Digital repositories are usually used by students, researchers, and other interested parties, with an objective to research a certain topic and broaden their knowledge in that domain. A recommender system can be very helpful in achieving that, since it helps discover relevant documents, while the user does not need to browse and review a large amount of documents.

Recommender systems in academic digital repositories are becoming prominent as the number of produced academic documents in electronic format grows. There are many types of documents present in academic digital repositories, including, but not limited to, undergraduate theses, postgraduate theses (master's theses and doctoral theses), journal articles, conference articles, workbooks, study books, manuals, collections of problems, course slides, and other teaching and research materials. In Slovenia, universities, colleges, other higher education institutions, and research institutions have joined efforts to form the Slovenian Open-Access Infrastructure where documents from all partners would be publicly available. Naturally, this also provides a framework for recommender systems as it is possible to recommend documents between different institutions. Another positive side effect of this is that researchers from different institutions that are in the same field of expertise can see the work of their colleagues more transparently, encouraging cooperation between them. With this goal in mind, a recommender system for the Slovenian Open-Access Infrastructure was designed as a part of the infrastructure to support the goals of the nationwide project. The novelty of this recommender system is that it is currently the only recommender system in Slovenia that includes all Slovenian universities and their electronic publications. In practice, over 200,000 electronic publications originating from any of the Slovenian universities can be recommended using our system.

This paper presents a cascade type hybrid recommender system which is implemented in the Slovenian Open-Access Infrastructure with the aim to serve relevant document recommendations across all digital libraries and institutional repositories which are currently included in the infrastructure. The second section briefly reviews related work. The third section presents the current state of the Slovenian Open Access Infrastructure. The inner workings and the architecture of our recommender system are presented in the fourth section. In the fifth section, we give details on the feedback collection analysis for our implemented hybrid recommender system using the digital repositories established within the Slovenian Open Access Infrastructure. The sixth section contains conclusions and ideas for further work.

2. Related Work

Document recommender systems can be applied in many practical scenarios. Specifically, for the scenario of document recommendations where the documents are news, Reference [1] demonstrates the use of recommendations for job postings, in Reference [2], cloud computing was used for recommendations and Reference [3] demonstrates a semantic web approach to recommending news. Many document recommender systems have been extensively covered by the research field especially for use with news. References [4,5] provide a survey of news recommendation systems. In [6], fuzzy logic is used to recommend news using content-based methods. Rich feedback is used to recommend news to users in [7], while Reference [8] compares information retrieval algorithms in news recommendation scenarios. In some cases, semantic approaches such as Wordnet are used to aid in semantic recommendations [9,10].

Research paper recommender systems are also prominent when it comes to document recommendations [11]. A tag-based research paper recommender system framework is presented in [12], and a similar tag-based approach was used in [13]. A collaborative filtering approach using contexts was used to recommend research papers in [14]. An extensive comparison of offline and online evaluation approaches of research paper recommender systems is presented in [15]. Specifically for digital repositories, several recommender systems have been developed. In [16], keyphrases were used as a basis for research paper recommendations and, in [17], a social bookmarking service CiteULike was used for recommendations. A recommender system specifically tailored for advising research publications as a part of digital libraries in a university environment was presented in [18]. Another study [19] introduces a Recommendation-as-a-Service (RaaS) platform used for recommendations in academia and its integration into the reference manager JabRef [20]. Similarly, CORE Recommender [21] was developed specifically for use in digital libraries and repositories. As shown in [22], such recommender systems have also been implemented in academic social networks, namely Mendeley.

When faced with researching, implementing, and maintaining recommender systems, challenges do occur. Some major challenges were outlined in [23]. These include data quality, the lack of appropriate data sets, choice of appropriate recommendation techniques, evaluation of recommendations, and even the number of recommended items. In addition to these challenges, we also encountered challenges while processing documents in the Slovenian language. Being a morphologically rich language, it is required to take different approaches to natural language processing when processing documents in Slovenian. Very little research has been done in recommending documents in the Slovenian language, mostly because there was very few structured datasets of documents in Slovenian. With the introduction of the Slovenian Open Access Infrastructure [24], this has improved greatly due to the creation of a large structured dataset, containing over 200,000 documents [25]. It features segmented metadata consisting of titles, abstracts, keywords as well as full-texts and other document metadata. From it, other datasets of the Slovenian language have formed [26,27], which allows for further research options not only in the research of recommender systems, but also other tasks in information retrieval and natural language processing, specific to the Slovenian language.

3. Overview of the Slovenian Open Access Infrastructure

The Slovenian Open Access Infrastructure was established in 2013 and has since enabled the interested parties in Slovenia (researchers, students, companies, and the public) access to the intellectual production of Slovenian educational and research organizations. Simultaneously, it has enabled the researchers to fulfill the requirements for open access to publications from publicly financed research. Structurally (Figure 1), the infrastructure consists of a national portal OpenScience.si [28], institutional repositories for each of the four Slovenian universities (Digital Library of University of Maribor (DLUM) [29], Repository of University of Ljubljana (RUL) [30], Repository of University of Primorska (RUP) [31], Repository of University of Nova Gorica (RUNG) [32]), a repository for research institutions (Digital Repository of Slovenian Research Organizations (DiRROS) [33]), and a repository for colleges and higher education institutions (ReVIS) [34]).

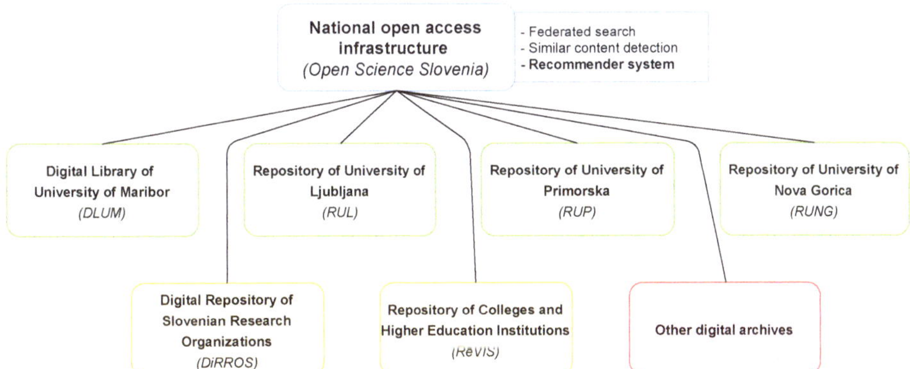

Figure 1. Structure of the Slovenian Open-Access Infrastructure.

The infrastructure also aggregates metadata from other digital archives such as videolectures.net [35], Social Science Data Archives [36], Digital Library of Slovenia [37], NUK Web Archive [38], and the Ministry of Defense Library and Information System [39]. The types of publications that are stored in the infrastructure include diploma, master's and doctoral theses, journal and conference articles, proceedings, datasets, scientific and technical reports, books, lecture materials, and videos of lectures. Since a great majority of publications are in Slovenian, a side product of this infrastructure was a large-scale corpus of full-text documents in the Slovenian language, covering several different domains of research. It also spawned some research datasets for use in linguistic studies [40,41]. More importantly, it currently represents the largest corpus of segmented texts in the Slovenian language, giving several options for research not only in linguistics but also in natural language processing. Due to interests for cooperation between the four universities and several research institutions in Slovenia, a recommender system was integrated in the infrastructure. The aim was to notify users about similar studies being done at different institutions through digital libraries and institutional repositories.

4. Document Recommendations

There are a few different approaches to recommendation in existence. The most common approaches are content-based and collaborative filtering [42,43]. Other approaches include demographic, utility-based, and knowledge-based techniques to recommendation. There is no optimal approach for every situation. Each approach has advantages and disadvantages in certain scenarios. While content-based filtering works well when a good description of an object is provided and when starting out with recommendations, collaborative filtering tends to provide more contextually appropriate recommendations once enough user feedback is provided. Hybrid systems aim to resolve the disadvantages of both approaches by combining them in different ways [44]. Several hybridization methods exist [45]. Weighted hybrids compute a score for a recommended item using outputs of all recommendation approaches available in the system. Switching hybrids employ a mechanism to switch between recommendation approaches. In this type of hybrid, approaches in the system are usually given priorities. If an approach with a higher priority cannot give a sufficient score, the recommender system switches to an approach with a lower priority as an attempt to provide a more recommendation with a more sufficient score. Mixed hybrids provide recommendations from different approaches at the same time. In cascade hybrids, one approach is used first to produce an initial set of recommended items; then, a second approach is used to fine-pick the most suitable items from that initial set, in order to provide a final recommendation.

Our recommender system is a cascade hybrid, incorporating content-based filtering as a primary recommendation technique and collaborative filtering as a secondary re-ranking method. It consists of three fundamental modules (Figure 2). The user activity log module provides the information on user activities such as view count, download count, document ratings, and document referrals. The document processing module ensures a unified feature representation of all documents in a triplet representation consisting of a title, keywords, and an abstract. Simultaneously, this module performs the calculation of BM25 values for each document pair, which forms a document index. The latter is a similarity matrix for all documents. Documents are periodically processed as new documents are added to the system daily. This way, the index is kept updated and the recommendations include new documents.

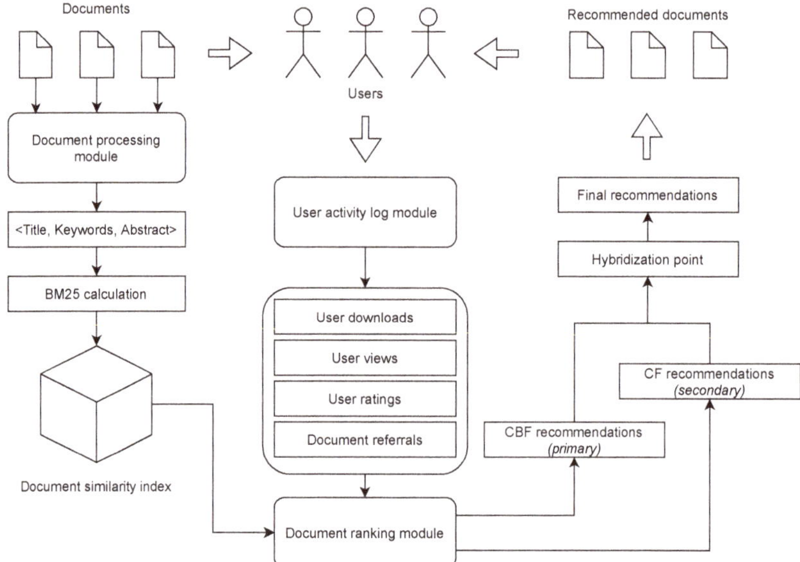

Figure 2. The architecture of our hybrid recommender system.

The user activity data and the calculated similarities between the documents are the input to the document ranking module, where similar documents are chosen depending on the document that is viewed by the user. This is also the hybridization point, where content-based filtering and collaborative filtering methods are applied in cascade to output the final list of recommendations, which is served to the end-user.

4.1. Processing Documents in Slovenian

A variety of different metadata were obtained from previous established repositories. These included information about authors, titles, keywords, abstracts, publishing year, and other bibliographic information. The metadata standards were different and included COMARC, MARC 21, and Dublin Core Metadata. We merged the different metadata schemes in our own metadata scheme to enable collection of as much metadata as possible. Our own metadata scheme consists of all metadata fields from the established standards with some extra fields for internal use. We use our metadata scheme to represent documents and use it with the recommender system as well as some other services within the Slovenian Open-Access Infrastructure.

For the recommender system, the documents are represented by titles, keywords, and abstracts. Most documents are in the Slovenian language; however, there are also documents in English, German, Italian, Croatian, and Hungarian. The documents that are not written in Slovenian have at least the abstract and keywords translated to Slovenian to conform with the publication and cataloguing rules. In the case of these documents, the available metadata in Slovenian are used with higher priority than the metadata in other languages. First, the most common words in the Slovenian language are removed from the text, since they do not contribute to semantic information. These are mainly conjunctions, prepositions, particles, and interjections; however, common verbs and nouns are also included. The common word list was built using word counts in documents. This is a periodic task, which is run each time after a recommendation index is updated. Additionally, we used lemmatization to help when dealing with conjugations and declensions in the text. Lemmatization is the process of determining the basic lexical form (i.e., lemma) to the words in a text. A very similar process to lemmatization is stemming. The main difference between lemmatization and stemming is that stemming does not convert the word into its dictionary form but simply cuts off the ending of the word. In text mining, lemmatization can be used to detect contexts of texts. It is used in our text processing step to group semantically similar words and to avoid the difficult process of grouping with declension and conjugation rules. Furthermore, n-grams for $N = [1, 2, 3, 4, 5]$ are generated and used with the tf-idf based ranking function BM25 to perform content-based filtering within our hybrid approach to recommendation.

4.2. Document Ranking

For document ranking, we used the BM25 ranking function [46] along with additional weights, which were obtained from document metadata and user activities. BM25 is a ranking function, which enables the ranking of documents by the similarity of terms that are contained within those documents. It is a family of functions, which differs by weighting schemes and parameter values. In general, tf and idf weights are used [47]. The term frequency (tf) is the occurrence count of a term t within a document d while the inverse document frequency (idf) is the importance of the term t in the given document collection D (Equation (1)). Composite nonlinear tf normalizations and the family of BM25 ranking functions have been used extensively in search engines to rank documents:

$$idf(t) = \log \frac{||D|| - n(t) + 0.5}{n(t) + 0.5} \tag{1}$$

$$s(d, Q) = \sum_{i=1}^{||Q||} idf(q_i) \cdot \frac{tf(q_i, d) \cdot (k_1 + 1)}{tf(q_i, d) + k_1 \cdot B}, \quad q_i \in Q, d \in D \tag{2}$$

$$B = 1 - b + b \cdot \frac{l_d}{avgdl} \tag{3}$$

It is a state-of-the-art tf-idf based ranking function and has spawned many variants including BM25L, BM25+, BM25-adpt and BM25T [48,49], which bring improvements on very specific datasets. It has also been implemented in open source and commercial solutions such as Apache Lucene, Apache Solr, and Xapian as well as in Microsoft SQL Server and MySQL database implementations as a default full-text search solution. We decided to implement BM25 ourselves on a Microsoft SQL Server platform to have research options while studying parameters of the original ranking function and its variants, since commercial solutions do not allow enough customization. Another reason for this is that our documents are in the Slovenian language, for which only limited support exists in these open source and commercial solutions.

$||D||$ in Equation (1) is the length of the collection D and $n(t)$ is the number of documents which contain the term t. The BM25 value $s(d,q)$ depends on the weights tf and idf as well as parameters k_1 and b. A general BM25 calculation for a document d and a query q with terms q_i is given with Equation (2), where $||Q||$ is the size of the query Q given with the number of terms and B is a normalization factor (Equation (3)). In Equation (3), l_d is the length of document d and $avgdl$ is the average length of the document in the corpus D.

The parameter k_1 regulates the importance of the tf weight and the parameter b regulates the importance of document length. The values for these two parameters can be set using advanced optimization approaches, but usually values $k_1 \in [1.2, 2.0]$ and $b = 0.75$ are used [50]. Currently, we use empirically determined fixed values $k_1 = 1.2$ and $b = 0.75$, but further study of the corpus properties and parameter effects is underway. An automated adaptive technique of choosing the parameters using an optimization method such as in [51] is desired. Additionally, we are also working on including alternative weighting schemes such as tf^*pdf [52] and tf-$id_u f$ [53].

4.3. Hybrid Approach to Recommendation

The input to our content-based filtering approach is a collection of metadata which describes the documents. A document feature is represented with a vector of terms obtained from titles, keywords, and abstracts. As we also have full-texts available, we empirically found that it is better to use semantically dense metadata rather than full-text due to two important disadvantages. Firstly, full-texts contain more terms which slows down the process of ranking similar documents. Secondly, semantically important contexts diminish even after applying pre-processing with stop-word lists and tf-idf filtering. However, when compared to a simpler document feature assembled from titles, keywords and abstracts do not significantly improve recommendation results. We further enrich the document feature with metadata including document typology [54], issue year, authors, repository ID, and document language.

With all the metadata considered, we calculate a BM25 score based on the enriched document features. We also use the Jaro–Winkler distance [55,56], in order to define a document typology similarity. The Jaro–Winkler similarity is suitable when dealing with short strings and when the similarity between them should be greater if the two strings match from the beginning. First, the Jaro similarity is calculated by including the number of matching characters m and half the number of transpositions t between strings s_1 and s_2 and their respective lengths $||s_i||$ (Equation (4)). Then, the Jaro–Winkler similarity is calculated by including the common prefix length λ and a scaling factor $p = 0.1$ to adjust the value depending on the common prefix length (Equation (5)). In our situation, the document typologies are denoted with a short string of up to five characters (e.g., $\lambda = 5$). The first character of the typology defines the kind of document and the following characters define the variant of the document. Some examples of document types are provided in Table 1.

Table 1. Examples of document typologies and their metadata notation. Full typology is available in [54].

Document Typology (Notation)	Document Typology (Meaning)
1.01	Original scientific article
1.02	Review article
1.03	Short scientific article
1.04	Professional article
2.08	Doctoral dissertation
2.09	Master's thesis
2.11	Undergraduate thesis
2.23	Patent application
2.24	Patent
2.25	Other monographs and completed works

Using the Jaro-Winkler distance (Equation (6)), we compare the typologies of two documents in order to rank the documents with the similar typology higher. The final content-based filtering score (Equation (7)) is calculated as a product between the BM25 score on the document feature vector and the Jaro–Winkler similarity on the document typology:

$$sim_j(s_1, s_2) = \begin{cases} 0 & \text{if } m = 0 \\ \frac{1}{3}\left(\frac{m}{|s_1|} + \frac{m}{|s_2|} + \frac{m-t}{m}\right) & \text{otherwise} \end{cases} \tag{4}$$

$$sim_{jw}(s_1, s_2) = sim_j(s_1, s_2) + \lambda p(1 - sim_j(s_1, s_2)) \tag{5}$$

$$d_{jw}(s_1, s_2) = 1 - sim_{jw}(s_1, s_2) \tag{6}$$

$$Score_{CBF} = BM25(d_A, d_B) \cdot d_{jw}(t_{d_A}, t_{d_B}) \tag{7}$$

Our collaborative filtering approach is collaborative in the sense that we use user interactions to re-rank the content-based filtering recommendations with the goal of improving recommendations. The input to our collaborative filtering approach is the user activity data regarding a document a_d. Views and download counts for documents are kept and regularly updated. The values for actions were set to 1 if a view occurs and 10 if a download occurs, meaning that a download action is as significant as 10 view actions (Equation (8)). A feedback value $f_{(a_d)}$ is calculated by summing all values of actions. Furthermore, we also store a similar feedback value for actions r_d on recommended documents $f_{(r_d)}$ to give higher weight to the documents which were interesting to end-users (Equation (9)). The values for boosts were set to 5 if a view on a recommended document occurs and 50 if a recommended document is downloaded. Action significance values for a_d and r_d were set empirically, with an idea in mind that a download is worth 10 times as significant as a view, and a recommended view is five times as significant as a regular view.

$$f_{a_d} = \sum_{i=1}^{||a_d||} a_{d,i} \qquad a_{d,i} = \begin{cases} 1 & \text{document view} \\ 10 & \text{document download} \end{cases} \tag{8}$$

$$f_{r_d} = \sum_{i=1}^{||r_d||} r_{d,i} \qquad r_{d,i} = \begin{cases} 5 & \text{document view} \\ 50 & \text{document download} \end{cases} \tag{9}$$

We can provide adaptive recommendations using actions from users by combining feedback values for actions and recommendations with the download rate h_d (Equation (10)), which is the ratio between downloads and views of a document. The logic is the same for the download rate of the recommended documents h_r, but only views and downloads on the recommended document are considered. The feedback value for actions on recommended documents makes the clicked recommendations rank higher in the recommendation list. The final collaborative filtering score (Equation (11)) is calculated as a product of the document download rate and the sum of action feedback values on the document and actions on recommendations:

$$h_d = \frac{\text{downloads}(d)}{\text{views}(d)} \qquad h_r = \frac{\text{downloads}(d_r)}{\text{views}(d_r)} \tag{10}$$

$$Score_{CF} = f_{a_d} \cdot h_d + f_{r_d} \cdot h_r \tag{11}$$

With both approaches combined into a hybrid approach, we use recommendation strategies, which can be customized depending on the type or purpose of recommendations. Some recommendation strategies that we used in production are »latest + relevant«, »same repository + relevant« and »more from same authors«. These strategies can also be merged into a single strategy using priority factors. For example, a strategy »latest from same repository and from same authors« would first pick the latest documents and would then filter them according to their repository primarily and according to their authors secondarily.:

$$\tau_d = \delta^{\text{Year}_{\text{now}} - \text{Year}_d} \tag{12}$$

The workflow of our hybrid recommender system consists of four steps (Figure 3). First, the results from our content-based approach are obtained. Second, an exponential temporal decay mechanic (Equation (12)) is implemented to increase the ranks of recently published documents. The parameter δ controls the exponential temporal decay. The similarity score of the document is multiplied by the temporal decay and the recommendations in the results are re-ranked. Documents contained in the result set are then input into our collaborative filtering approach which re-ranks the results again. Currently, the output result length of our content-based approach is 25 documents. Finally, the list of recommendations is shortened to N documents for better presentation of the result on the web. In practice, we shorten the list to five documents.

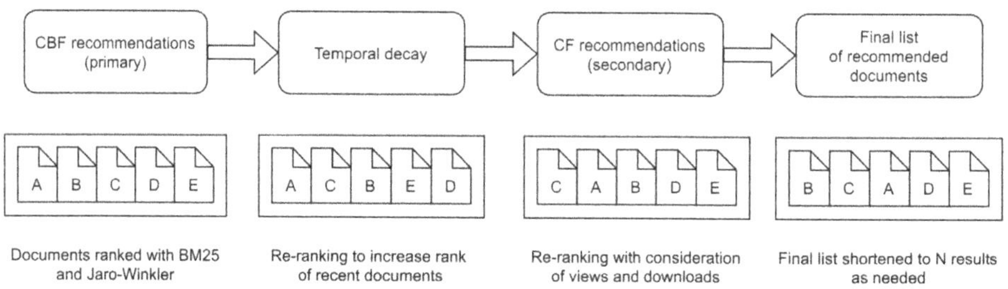

Figure 3. The workflow of our hybrid cascade approach to recommendations.

5. Feedback Collection Analysis

Collecting feedback from users is an important part of recommender systems design because it can directly influence the resulting recommendations. The overall user experience with regard to recommendations can be greatly improved if feedback is regularly collected from users. This can be done directly using surveys, questionnaires, and quick questions or indirectly by analyzing user activity. To achieve sufficient feedback, an appropriate time for feedback collection must be determined. The quality of feedback depends on the mood of the user, but, with careful planning, there is more chance that the user will be willing to give good quality feedback. Another perspective is to collect feedback at a certain time, where we are sure that users might be more inclined to express their opinions (e.g., a week after something changed) as they have had enough time to form an opinion. Furthermore, a good feedback collection approach can lead to an organized approach to evaluation of recommender systems. With it, evaluation metrics can be better defined and used to measure the true performance of the recommender system.

We performed an analysis of time-frames during the year, when feedback collection would make sense within the Slovenian Open-Access Infrastructure. In our case, the recommendations are focused on documents and are meant to help students, academic staff, and researchers find more similar documents to their interest. The recommendations are therefore accessed as the users are using the recommender system, which is linked to different time-frames during the year. We found that several spikes in usage occur during the year and we tried to link them to specific events that occur in the academic year (e.g., thesis defenses, summer vacations, etc.).

We limited our data to data from four universities in Slovenia and their institutional repositories in the Slovenian Open-Access Infrastructure. University of Maribor was included with DLUM, University of Ljubljana with RUL, University of Primorska with RUP and University of Nova Gorica with RUNG.

All institutional repositories store view and download counts for documents. During this analysis, we treated viewed documents as mildly interesting and downloaded documents as very interesting. We did this because a download can occur only after the document is viewed; therefore, if a user downloaded the document, they must have viewed its detailed description with metadata and made a conscious decision that it is interesting enough for them to download it.

We encountered a major limitation with the accessibility of the traffic data on each institutional repository. DLUM was the only repository that we were able to get the data from, since other repositories opted not to be included in the analysis by their maintenance teams. Furthermore, the maintenance teams of DLUM, RUL, RUP, and RUNG decided to exclude all traffic tracking options on repositories after 2016. As for DLUM data that we were able to obtain, it was Google Analytics traffic data between January 2013 and December 2016. With all limitations considered, we performed an analysis using data only from DLUM (Figure 4). It proved to be a suitable institutional repository for this task, since it is the first university institutional repository in Slovenia, running since 2008 and serving as a basis for all other institutional repositories in the national open-access infrastructure.

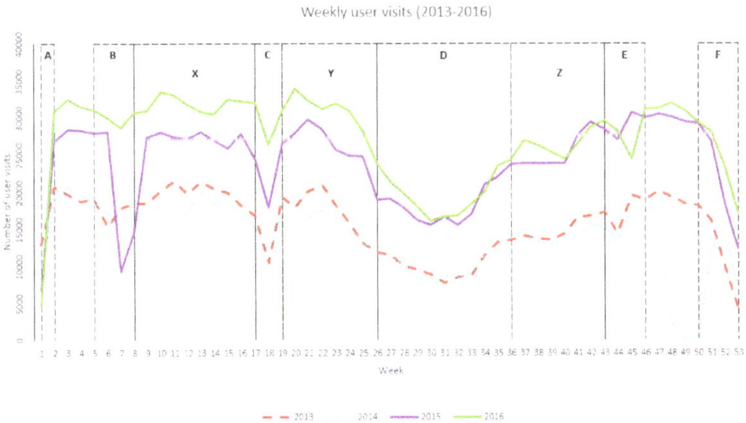

Figure 4. Weekly user visits to DLUM between January 2013 and December 2016.

In the data set time frame of user activity between January 2013 and December 2016 (Figure 4), special events have occurred. In November 2014, DLUM saw a major update and was offline for two weeks (weeks 48 to 50) due to this. It was updated at this time because it had to run stable for most of the year, due to a regular influx of new theses. This influx annually reaches a peak in September and October (weeks 40 to 42), when the theses are catalogued by the librarians. It was decided to run

DLUM without interruption between March and November 2014 because most users during that time are students researching for their theses and researchers searching for related work for their articles.

An increase in weekly user visits can be observed in 2015. This increase seems to be attributed to the marketing efforts of the Slovenian Open-Access Infrastructure and the cross-repository recommendations; however, this cannot be confirmed due to the lack of traffic tracking capabilities on repositories RUL, RUP, and RUNG.

Furthermore, in 2016, we can observe another increase in weekly user visits, which lasts from January (week 1) to September (week 40). This unusual additional traffic was generated by students enrolled in pre-Bologna process study programs at the University of Maribor. These students had to complete and defend their theses by October 2016 as directed by the University of Maribor and were most likely collecting research on DLUM in order to achieve this. This reason holds, as the traffic increase stops in September 2016 (week 40).

By observing traffic fluctuation during the year, we found a decrease in weeks that correspond to holidays. This occurs in several time-frames which are visible in Figure 4 and denoted with letters:

- A—January; the first week of the year (consequence of New Year),
- B—February; weeks 7 and 8, around February 8th (national holiday "Prešeren Day"),
- C—April and May; week 18 and 19, starting around April 27th (national holiday "Day of uprising against occupation") and ending around May 1st (national holiday "International Workers' Day"),
- D—June, July and August; weeks 26 to 36, summer holiday season,
- E—October, November; weeks 44 and 45, around October 31st (national holiday "Reformation Day") and November 1st (national holiday "All Saint's Day"),
- F—December; weeks 50 to 53, around December 25th (national holiday "Christmas"), 26th (national holiday "Independence and Unity Day") and December 31st (national holiday "New Year's Eve").

We conclude that these time-frames are suitable for maintenance work on institutional repositories. Time-frames B, C, and E show the potential for smaller updates and minor changes, while time-frame D shows the potential for large-scale maintenance.

We also observed the peak traffic occurring between some before mentioned time-frames:

- X—weeks 9 and 17 (from February to April),
- Y—weeks 20 to 25 (from May to June),
- Z—weeks 37 to 43 (from August to October).

We conclude that these time-frames are suitable for feedback collection campaigns, surveys, and questionnaires. Namely, time-frames X and Y are more suitable for active user feedback collection (e.g., validation of recommended documents), since users are actively researching during that time. Time-frame Z is more suitable for general feedback collection (e.g., general surveys regarding user experience).

An extensive evaluation study of our recommender system is currently still underway as it requires successful collaboration of several institutions that maintain their own repositories. Several metrics for recommendation system evaluation exist. In general, there are two ways of evaluating any recommendation system: online and offline [15,57,58]. Offline evaluation makes use of preferably labelled data which is split into training and test sets. The recommendation system uses the training set ratings to try and predict the ratings in the test set. Actual users are not needed in this type of evaluation. This makes offline evaluation fast and easy to perform on a large amount of data. It can also be performed using many different datasets and with multiple different algorithms. The main disadvantage of this approach is that it cannot measure true user satisfaction.

In an online evaluation scenario, users interact with a running recommendation system and respond to it naturally, while feedback is being collected from them. Feedback is obtained by either asking the users directly or observing their actions. This approach measures true user satisfaction but can take a long time to set-up and run from beginning to end.

The choice of metrics differs depending on the approach of recommendation. Information retrieval metrics such as accuracy, recall, precision, and F-measure are usually considered preferable when evaluating content-based recommendation systems. Other metrics for this type of recommendation system include normalized discounted cumulative gain [59], rank-biased precision [60], and expected reciprocal rank [61]. Collaborative filtering recommendations are usually evaluated using approaches that measure novelty, serendipity, diversity, and coverage [62]. Currently, there are several different metrics [63] that can be used to evaluate recommendation systems. When dealing with hybrid recommendation systems, this must be carefully considered, since the type of hybridization can also affect the evaluation process, making it complex due to implementation in multiple stages.

6. Conclusions

In this article, we present a cascade hybrid recommender system implemented in institutional repositories that is part of the Slovenian National Open-Access Infrastructure. We outlined the recommender system architecture, document pre-processing, and ranking approaches. A feedback collection analysis has been presented on real-world data from one of our longest running repositories. With the analysis, we were able to identify different time-frames during the year where it is suitable to consider feedback collection on an academic digital repository. An extensive evaluation study is currently underway and we conclude that, for an extensive evaluation of our recommender system's contribution to knowledge exchange and spread across the Slovenian Open-Access Infrastructure, a unified framework should be developed in addition to institutional repository management processes regarding logging user activities and using traffic tracking scripts. Only with such an approach can a definitive contribution of the recommender system be confirmed and further researched. It would also allow the observation of any significant cooperation between institutions, as it is already suspected that the institutions in the two largest institutional repositories in the national open-access infrastructure be in accordance with the majority of research cooperation efforts in Slovenia.

Author Contributions: Conceptualization, M.B., M.F., and M.O.; methodology, M.B. and M.F.; software, M.B., M.F., and J.B.; validation, M.B., M.F., J.B., S.M., K.K., and M.O.; writing—original draft preparation, M.B. and M.O.; writing—review and editing, M.B., M.F., J.B., S.M., K.K., and M.O. All authors have read and agreed to the published version of the manuscript.

Funding: This research received no external funding.

Conflicts of Interest: The authors declare no conflict of interest.

References

1. Elsafty, A.; Riedl, M.; Biemann, C. Document-based Recommender System for Job Postings using Dense Representations. In Proceedings of the 2018 Conference of the North American Chapter of the Association for Computational Linguistics: Human Language Technologies, New Orleans, LA, USA, 1–6 June 2018; Volume 3, pp. 216–224. [CrossRef]
2. del Campo, J.V.; Pegueroles, J.; Hernández-Serrano, J.; Soriano, M. DocCloud: A document recommender system on cloud computing with plausible deniability. *Inf. Sci.* **2014**, *258*, 387–402. [CrossRef]
3. Cantador, I.; Bellogín, A.; Castells, P. News@hand: A Semantic Web Approach to Recommending News. In *International Conference on Adaptive Hypermedia and Adaptive Web-Based Systems, Proceedings of the AH 2008: Adaptive Hypermedia and Adaptive Web-Based Systems, Hannover, Germany, 29 July–1 August 2008*; Nejdl, W., Kay, J., Pu, P., Herder, E., Eds.; Springer: Berlin/Heidelberg, Germany, 2008; pp. 279–283.
4. Karimi, M.; Jannach, D.; Jugovac, M. News recommender systems—Survey and roads ahead. *Inf. Process. Manag.* **2018**, *54*, 1203–1227. [CrossRef]
5. Borges, H.L.; Lorena, A.C. A Survey on Recommender Systems for News Data. In *Smart Information and Knowledge Management: Advances, Challenges, and Critical Issues*; Szczerbicki, E., Nguyen, N.T., Eds.; Springer: Berlin/Heidelberg, Germany, 2010; pp. 129–151. [CrossRef]

6. Adnan, M.N.M.; Chowdury, M.R.; Taz, I.; Ahmed, T.; Rahman, R.M. Content based news recommendation system based on fuzzy logic. In Proceedings of the 2014 International Conference on Informatics, Electronics Vision (ICIEV), Dhaka, Bangladesh, 23–24 May 2014; pp. 1–6.

7. Ardissono, L.; Petrone, G.; Vigliaturo, F. News Recommender Based on Rich Feedback. In *International Conference on User Modeling, Adaptation, and Personalization, Proceedings of the UMAP 2015: User Modeling, Adaptation and Personalization, Dublin, Ireland, 29 June–3 July 2015*; Ricci, F., Bontcheva, K., Conlan, O., Lawless, S., Eds.; Springer International Publishing: Cham, Switzerland, 2015; pp. 331–336.

8. Bogers, T.; van den Bosch, A. Comparing and Evaluating Information Retrieval Algorithms for News Recommendation. In Proceedings of the 2007 ACM Conference on Recommender Systems (RecSys '07), Minneapolis, MN, USA, 19–20 October 2007; pp. 141–144. [CrossRef]

9. Capelle, M.; Hogenboom, F.; Hogenboom, A.; Frasincar, F. Semantic News Recommendation Using Wordnet and Bing Similarities. In Proceedings of the 28th Annual ACM Symposium on Applied Computing (SAC '13), Coimbra, Portugal, 18–22 March 2013; Association for Computing Machinery: New York, NY, USA, 2013; pp. 296–302. [CrossRef]

10. Capelle, M.; Moerland, M.; Hogenboom, F.; Frasincar, F.; Vandic, D. Bing-SF-IDF+: A Hybrid Semantics-Driven News Recommender. In Proceedings of the 30th Annual ACM Symposium on Applied Computing (SAC '15), Salamanca, Spain, 13–17 April 2017; pp. 732–739. [CrossRef]

11. Beel, J.; Gipp, B.; Langer, S.; Breitinger, C. Research-paper recommender systems: A literature survey. *Int. J. Digit. Libr.* **2016**, *17*, 305–338. [CrossRef]

12. Jomsri, P.; Sanguansintukul, S.; Choochaiwattana, W. A Framework for Tag-Based Research Paper Recommender System: An IR Approach. In Proceedings of the 2010 IEEE 24th International Conference on Advanced Information Networking and Applications Workshops, Perth, WA, Australia, 20–23 April 2010; pp. 103–108.

13. Choochaiwattana, W. Usage of tagging for research paper recommendation. In Proceedings of the 2010 3rd International Conference on Advanced Computer Theory and Engineering (ICACTE), Chengdu, China, 20–22 August 2010; Volume 2, pp. V2-439–V2-442.

14. Winoto, P.; Tang, T.; McCalla, G. Contexts in a Paper Recommendation System with Collaborative Filtering. *Int. Rev. Res. Open Distance Learn.* **2012**, *13*, 56–75. [CrossRef]

15. Beel, J.; Langer, S. A Comparison of Offline Evaluations, Online Evaluations, and User Studies in the Context of Research-Paper Recommender Systems. In *International Conference on Theory and Practice of Digital Libraries, Proceedings of the TPDL 2015: Research and Advanced Technology for Digital Libraries, Poznan, Poland, 14–18 September 2015*; Kapidakis, S., Mazurek, C., Werla, M., Eds.; Springer International Publishing: Cham, Switzerland, 2015; pp. 153–168.

16. Ferrara, F.; Pudota, N.; Tasso, C. A Keyphrase-Based Paper Recommender System. In *Digital Libraries and Archives*; Agosti, M., Esposito, F., Meghini, C., Orio, N., Eds.; Springer: Berlin/Heidelberg, Germany, 2011; pp. 14–25.

17. Bogers, T.; van den Bosch, A. Recommending Scientific Articles Using Citeulike. In Proceedings of the 2008 ACM Conference on Recommender Systems (RecSys '08), Lausanne, Switzerland, 23–25 October 2008; pp. 287–290. [CrossRef]

18. Porcel, C.; Moreno, J.; Herrera-Viedma, E. A multi-disciplinar recommender system to advice research resources in University Digital Libraries. *Expert Syst. Appl.* **2009**, *36*, 12520–12528. [CrossRef]

19. Beel, J.; Aizawa, A.; Breitinger, C.; Gipp, B. Mr. DLib: Recommendations-as-a-Service (RaaS) for Academia. In Proceedings of the 2017 ACM/IEEE Joint Conference on Digital Libraries (JCDL), Toronto, ON, Canada, 19–23 June 2017; pp. 1–2. [CrossRef]

20. Feyer, S.; Siebert, S.; Gipp, B.; Aizawa, A.; Beel, J. Integration of the Scientific Recommender System Mr. DLib into the Reference Manager JabRef. In *European Conference on Information Retrieval, Proceedings of the ECIR 2017: Advances in Information Retrieval, Aberdeen, UK, 8–13 April 2017*; Springer: Cham, Switzerland, 2017. [CrossRef]

21. Knoth, P.; Anastasiou, L.; Charalampous, A.; Cancellieri, M.; Pearce, S.; Pontika, N.; Bayer, V. Towards effective research recommender systems for repositories. *arXiv* **2017**, arXiv:1705.00578.

22. Vargas, S.; Hristakeva, M.; Jack, K. Mendeley: Recommendations for Researchers. In Proceedings of the 10th ACM Conference on Recommender Systems (RecSys '16), Boston, MA, USA, 15–19 September 2016; p. 365.

23. Beel, J.; Dinesh, S. Real-World Recommender Systems for Academia: The Pain and Gain in Building, Operating, and Researching them [Long Version]. *arXiv* **2017**, arXiv:1704.00156.

24. Ojsteršek, M.; Brezovnik, J.; Kotar, M.; Ferme, M.; Hrovat, G.; Bregant, A.; Borovič, M. Establishing of a Slovenian open access infrastructure: A technical point of view. *Program* **2014**, *48*, 394–412. [CrossRef]

25. OpenScience Slovenia Dataset. Available online: http://www.openscience.si/OpenData.aspx (accessed on 23 October 2020).

26. Erjavec, T.; Fišer, D.; Ljubešić, N.; Arhar Holdt, Š.; Bren, U.; Robnik Šikonja, M.; Udovič, B. Terminology Identification Dataset KAS-Term 1.0. Available online: https://www.clarin.si/repository/xmlui/handle/11356/1198 (accessed on 23 October 2020).

27. Erjavec, T.; Fišer, D.; Ljubešić, N.; Bitenc, M. Bilingual Terminology Extraction Dataset KAS-Biterm 1.0. Available online: https://www.clarin.si/repository/xmlui/handle/11356/1199 (accessed on 23 October 2020).

28. OpenScience Slovenia. Available online: https://www.openscience.si/ (accessed on 23 October 2020).

29. Digital Library of University of Maribor-DLUM. Available online: https://dk.um.si/info/index.php/eng (accessed on 23 October 2020).

30. Repository of the University of Ljubljana-RUL. Available online: https://repozitorij.uni-lj.si/info/index.php/eng (accessed on 23 October 2020).

31. Repository of the University of Primorska-RUP. Available online: https://repozitorij.upr/info/index.php/eng (accessed on 23 October 2020).

32. Repository of the University of Nova Gorica-RUNG. Available online: https://repozitorij.ung.si/info/index.php/eng (accessed on 23 October 2020).

33. Digital repository of Slovenian Research Organizations. Available online: https://dirros.openscience.si/info/index.php/eng (accessed on 23 October 2020).

34. Repository of Colleges and Higher Education Institutions-ReVIS. Available online: https://revis.openscience.si/info/index.php/eng (accessed on 23 October 2020).

35. Videolectures.net. Available online: https://videolectures.net (accessed on 23 October 2020).

36. Social Science Data Archives. Available online: https://www.adp.fdv.uni-lj.si/eng/ (accessed on 23 October 2020).

37. Digital Library of Slovenia. Available online: http://dlib.si/?=&language=eng (accessed on 23 October 2020).

38. NUK Web Archive. Available online: http://arhiv.nuk.uni-lj.si (accessed on 23 October 2020).

39. Ministry of Defence Library and Information System. Available online: https://dk.mors.si/info/index.php/en (accessed on 23 October 2020).

40. Jakubíček, M.; Fiser, D.; Suchomel, V. Terminology Extraction for Academic Slovene Using Sketch Engine. In Proceedings of the Tenth Workshop on Recent Advances in Slavonic Natural Language Processing, Karlova Studanka, Czech Republic, 2–4 December 2016; Volume 10.

41. Ljubešić, N.; Fiser, D.; Erjavec, T. KAS-term: Extracting Slovene Terms from Doctoral Theses via Supervised Machine Learning. In *International Conference on Text, Speech, and Dialogue, Proceedings of the TSD 2019: Text, Speech, and Dialogue, Ljubljana, Slovenia, 11–13 September 2019*; Springer: Cham, Switzerland, 2019; pp. 115–126. [CrossRef]

42. Adomavicius, G.; Tuzhilin, A. Toward the next generation of recommender systems: A survey of the state-of-the-art and possible extensions. *IEEE Trans. Knowl. Data Eng.* **2005**, *17*, 734–749. [CrossRef]

43. Bobadilla, J.; Ortega, F.; Hernando, A.; Gutiérrez, A. Recommender systems survey. *Knowl.-Based Syst.* **2013**, *46*, 109–132. [CrossRef]

44. Burke, R. Hybrid Recommender Systems: Survey and Experiments. *User Model. User-Adapt. Interact.* **2002**, *12*, 331–370. [CrossRef]

45. Burke, R. Hybrid Web Recommender Systems. In *The Adaptive Web: Methods and Strategies of Web Personalization*; Brusilovsky, P., Kobsa, A., Nejdl, W., Eds.; Springer: Berlin/Heidelberg, Germany, 2007; pp. 377–408. [CrossRef]

46. Robertson, S.; Zaragoza, H. The Probabilistic Relevance Framework: BM25 and beyond. *Found. Trends Inf. Retr.* **2009**, *3*, 333–389. [CrossRef]

47. Jones, K.; Walker, S.; Robertson, S. A probabilistic model of information retrieval: Development and comparative experiments: Part 2. *Inf. Process. Manag.* **2000**, *36*, 809–840. [CrossRef]

48. Géry, M.; Largeron, C. BM25t: A BM25 extension for focused information retrieval. *Knowl. Inf. Syst.* **2012**, *32*, 217–241. [CrossRef]

49. Trotman, A.; Puurula, A.; Burgess, B. Improvements to BM25 and Language Models Examined. In Proceedings of the 2014 Australasian Document Computing Symposium (ADCS '14), Melbourne, VIC, Australia, 27–28 November 2014; ACM: New York, NY, USA, 2014; pp. 58–65. [CrossRef]

50. Manning, C.D.; Raghavan, P.; Schütze, H. In *Introduction to Information Retrieval*; Cambridge University Press: New York, NY, USA, 2008.

51. Bollegala, D.; Noman, N.; Iba, H. RankDE: Learning a Ranking Function for Information Retrieval Using Differential Evolution. In Proceedings of the 13th Annual Conference on Genetic and Evolutionary Computation (GECCO '11), Dublin, Ireland, 12–16 July 2011; pp. 1771–1778. [CrossRef]

52. Nguyen, K.; Shin, B.-J.; Yoo, S.J. Hot topic detection and technology trend tracking for patents utilizing term frequency and proportional document frequency and semantic information. In Proceedings of the 2016 International Conference on Big Data and Smart Computing (BigComp), Hong Kong, China, 18–20 January 2016; pp. 223–230.

53. Beel, J.; Langer, S.; Gipp, B. TF-IDuF: A Novel Term-Weighting Scheme for User Modeling based on Users' Personal Document Collections. In Proceedings of the iConference 2017, Wuhan, China, 22–25 March 2017; doi: 10.9776/17217. [CrossRef]

54. COBISS/IZUM, Typology of Documents/Works for Bibliography Management in COBISS. 2016. Available online: https://home.izum.si/COBISS/bibliografije/Tipologija_eng.pdf (accessed on 23 October 2020).

55. Jaro, M.A. Advances in Record-Linkage Methodology as Applied to Matching the 1985 Census of Tampa, Florida. *J. Am. Stat. Assoc.* **1989**, *84*, 414–420. [CrossRef]

56. Winkler, W. String Comparator Metrics and Enhanced Decision Rules in the Fellegi-Sunter Model of Record Linkage. Available online: https://files.eric.ed.gov/fulltext/ED325505.pdf (accessed on 23 October 2020).

57. Hernández del Olmo, F.; Gaudioso, E. Evaluation of recommender systems: A new approach. *Expert Syst. Appl.* **2008**, *35*, 790–804. [CrossRef]

58. Silveira, T.; Zhang, M.; Lin, X.; Liu, Y.; Ma, S. How good your recommender system is? A survey on evaluations in recommendation. *Int. J. Mach. Learn. Cybern.* **2019**, *10*, 813–831. [CrossRef]

59. Wang, Y.; Wang, L.; Li, Y.; He, D.; Liu, T.Y. A Theoretical Analysis of NDCG Type Ranking Measures. In *Conference on Learning Theory*; PMLR: Princeton, NJ, USA, 2013; Volume 30, pp. 25–54.

60. Moffat, A.; Zobel, J. Rank-Biased Precision for Measurement of Retrieval Effectiveness. *ACM Trans. Inf. Syst.* **2008**, *27*. [CrossRef]

61. Chapelle, O.; Metlzer, D.; Zhang, Y.; Grinspan, P. Expected Reciprocal Rank for Graded Relevance. In Proceedings of the 18th ACM Conference on Information and Knowledge Management (CIKM '09), Hong Kong, China, 2–6 November 2009; Association for Computing Machinery: New York, NY, USA, 2009; pp. 621–630. [CrossRef]

62. Gunawardana, A.; Shani, G. A Survey of Accuracy Evaluation Metrics of Recommendation Tasks. *J. Mach. Learn. Res.* **2009**, *10*, 2935–2962.

63. Shani, G.; Gunawardana, A. Evaluating Recommendation Systems. In *Recommender Systems Handbook*; Springer: Boston, MA, USA, 2011; pp. 257–297. [CrossRef]

 *information*

Article

Graph Convolutional Neural Network for a Pharmacy Cross-Selling Recommender System

Franz Hell [1,*,†], Yasser Taha [1,†], Gereon Hinz [1,*], Sabine Heibei [2], Harald Müller [2] and Alois Knoll [3]

1. STTech GmbH, 80939 Munich, Germany; ysr.mahmoud@gmail.com
2. Pharmatechnik, 82319 Starnberg, Germany; s.heibei@pharmatechnik.de (S.H.); h.mueller@pharmatechnik.de (H.M.)
3. Munich School of Robotics and Machine Intelligence, Technical University of Munich, 80333 Munich, Germany; knoll@mytum.de
* Correspondence: franz.hell@sttech.de (F.H.); gereon.hinz@sttech.de (G.H.)
† These authors contributed equally to this work.

Received: 29 September 2020; Accepted: 8 November 2020; Published: 11 November 2020

Abstract: Recent advancements in deep neural networks for graph-structured data have led to state-of-the-art performance in recommender system benchmarks. Adapting these methods to pharmacy product cross-selling recommendation tasks with a million products and hundreds of millions of sales remains a challenge, due to the intricate medical and legal properties of pharmaceutical data. To tackle this challenge, we developed a graph convolutional network (GCN) algorithm called PharmaSage, which uses graph convolutions to generate embeddings for pharmacy products, which are then used in a downstream recommendation task. In the underlying graph, we incorporate both cross-sales information from the sales transaction within the graph structure, as well as product information as node features. Via modifications to the sampling involved in the network optimization process, we address a common phenomenon in recommender systems, the so-called popularity bias: popular products are frequently recommended, while less popular items are often neglected and recommended seldomly or not at all. We deployed PharmaSage using real-world sales data and trained it on 700,000 articles represented as nodes in a graph with edges between nodes representing approximately 100 million sales transactions. By exploiting the pharmaceutical product properties, such as their indications, ingredients, and adverse effects, and combining these with large sales histories, we achieved better results than with a purely statistics based approach. To our knowledge, this is the first application of deep graph embeddings for pharmacy product cross-selling recommendation at this scale to date.

Keywords: graph convolutional neural network; recommender system; cross-sales; pharmacy; popularity bias

---

1. Introduction

Deep learning algorithms play an increasingly important role in recommender systems. In recent years, new deep learning architectures known as graph convolutional networks (GCNs) that can learn from graph-structured data [1–4] were introduced and applied to recommendation applications. The basic principle behind GCNs is to use neural networks to learn how to iteratively aggregate and transform feature information from a local graph neighborhood to obtain a final representation of a given node, called the "embedding". This way, GCNs can incorporate both feature information, as well as the graph structure. These methods can be leveraged to distill useful low-dimensional embeddings of input data such as images, text, molecules, or individual users [5–8]. These low-dimensional embeddings then can be used in downstream tasks such as recommendation applications, where they can complement or even replace traditional recommendation algorithms, such as collaborative filtering. GCN based methods have set a new standard on countless recommender system benchmarks [1] and are now being used in real-world production environments such as Pinterest.

1.1. Present Work

The current state of cross-selling in pharmacies relies heavily on expert knowledge, which is typically provided by salespeople interacting directly with customers. The feedback loop for improving the knowledge about possible cross-sales is therefore very localized, with a very poor distribution of experience gained. Our recommender system has the goal of substantially improving the collection and distribution of general knowledge about possible cross-sales to pharmacy businesses, while taking into account the restrictions explained below. The task of our system is to recommend pharmaceutical products that are good cross-sales for a given pharmaceutical product. A good recommendation for a given query for example is an over-the-counter drug that is used for complementary therapy that the query product is intended for or helps with additional symptoms that the query product indications are associated with or a product that helps alleviate possible adverse effects of the query product. As sales of medical products are highly regulated in all countries, prescription medications can only be sold if prescribed by a physician, and hence are not legitimate cross-selling items.

The prescription pain killer medicine diclofenac is associated with the possible adverse reaction of increases in gastric acid, when taken for a long period of time. Pantoprazole, a proton-pump-inhibitor, can help alleviate this adverse effect by prohibiting excess production of stomach acid. Other good recommendations for diclofenac are additional gels for the treatment of blunt injuries such as contusions or sports injuries for which diclofenac is often prescribed, which mitigate swelling and additionally help in local pain therapy.

The setting for our recommendation system and experimental evaluation is the European Union, where the GDPR forbids, among other things, the processing of medical data for pure sales purposes if these data can be linked to specific persons, even if they are pseudonymized, as such information could allow for an individual data subject to be singled out and linked across different datasets. Traditional recommendation methods like collaborative filtering or content based approaches all require a user history (either user ratings or user-article interaction histories with information about the articles) for computing recommendations. A pharmacy cross-selling recommender system however is restricted from using the purchase history information. All our input data are therefore completely anonymized, and the only historical information that can be exploited is the co-occurrence of articles in sales transactions. The second data source we use is the description of indications, active ingredients, and adverse effects for each article.

This paper describes the main challenges when applying the idea of a recommender system based on graph convolutional networks to pharmaceutical product cross-selling recommendations, given the above-mentioned restrictions. The first challenge is to devise a recommendation system based on product feature information and sales data that solely represents which products have been sold together with no information about customer product relations. To address this challenge, we construct a graph representation of pharmacy sales and product data and employ a graph neural network to learn product embeddings, which can be used for a downstream recommendation task. We describe how we chose to encode this information into a graph structure by representing pharmaceutical products as nodes, their features as node features, and cross-sales-statistics as node-node edges.

We also describe the optimization goal of the graph neural network training process with respect to these data. We chose to use a semi-supervised training scheme by utilizing the triplet loss to optimize the network parameters. The definition of positive and negative samples is a key decision when using triplet loss in optimization. Here, we describe how we selected positive and negative samples based on cross-selling and feature information present in the graph.

A major problem when using sales statistics is the distribution of the sales data, which often shows that popular articles have a high probability of being sold, while the so-called long tail or distant tail [9] contains articles that have a low overall sales probability. This popularity bias also influences "cross-selling statistics", meaning that products that are sold often also dominate cross-selling statistics, even when they are not particularly related to the product with which they are sold. Here, we describe how we counteract this bias with a re-ranking approach based on probability theory and how we additionally use feature information in a hybrid sampling approach.

Finally, we describe how we evaluate the results of the algorithm by expert review of a selected segment of the generated cross-selling recommendations. In particular, we compare the results to purely statistics based cross-selling recommendations computed from the input dataset.

1.2. Related Work

1.2.1. Pharmaceutical Product Cross-Sales Recommendations

A study by Rutter et al. [10] interviewed sixteen pharmacists and found that pharmacists relied mostly on personal judgment augmented by patient feedback to make product recommendations. Another study examining the factors affecting pharmacists' recommendations of complementary medicines [11] also reported that recommendations are made based on personal experience and education and concluded that in order to encourage the informed use of complementary medicines in pharmacies, there is a need for the development of accessible, quality resources. A study investigating the community pharmacists' recommendations for alternative natural products for stress in Melbourne reported that out of 94 pharmacies, twenty-five provided the customer with an inappropriate product and concluded that there is a need to develop guidelines for pharmacists to make evidence based decisions in recommending complementary and alternative medications [12]. A study investigating a recommender system for newly marketed pharmaceutical drugs developed an adverse drug reactions label prediction component, which emits, for a given prescription, a set of drugs that should be avoided as they will induce adverse drug reactions if taken together with the prescription [13]. Zhang et al. [14] proposed a cloud-assisted drug recommendation system for online pharmacies to recommend users the top-N related medicines according to the symptoms. In their system, they first clustered the drugs into several groups according to the functional description information and designed a basic personalized drug recommendation based on user collaborative filtering. In the collaborative filtering based part of their pipeline, the authors leveraged drug user ratings. In our approach, we do not have access to historical user data or user ratings, as explained above. We can base our modeling solely on raw sales data and product features. Therefore, we designed a unique approach to solve the pharmaceutical product-product recommendation task based on node embeddings generated by a graph neural network that can leverage sales data, as well as product features without the need for customer histories or user ratings.

1.2.2. Graph Convolutional Neural Networks and Recommender Systems

Conceptually, our approach is related to previous node embedding algorithms and contemporary advancements in the application of convolutional neural networks to graph-structured data. The core idea behind node embedding methods is to create useful low-dimensional vector representations of high-dimensional information about a node in the graph, including the node's (local) graph neighborhood. The use of low-dimensional vector embeddings as feature inputs for a wide variety of machine learning tasks such as classification, prediction, clustering, and recommendation tasks has been proven to be valuable [4,15,16]. The original work of [17], which introduced a variant of graph convolutions that were based on spectral graph theory, was followed by several authors, proposing improvements and extensions [1–4,7,15,18–21]. Early approaches to generating node embeddings, such as the GCN introduced by [2], were transductive and limited in their generalization to unseen nodes and in their scalability since those methods required operating on the Laplacian of the entire graph [22]. Later approaches operated in an inductive fashion [23]. In contrast to transductive embedding methods that are based on matrix factorization, newer approaches leverage node features and the spatial structure of each node's local neighborhood, as well as the distribution of node features in the neighborhood [24]. Building on this, further improvements and new algorithmic features have recently been explored that ensure performance, scalability, and improved sampling [22,25,26]. These advancements led to new improved performance on benchmarks such as node classification, link prediction, or web-scale recommendation tasks, as well as the application of those methods to areas such as drug design [1,3,7,8,18].

1.2.3. Popularity Bias

One obstacle to the effectiveness of recommender systems is the problem of popularity bias [27]. Collaborative filtering based recommenders typically emphasize popular articles (e.g., those with more sales, views, or ratings) over articles from the long tail [28] that may only be popular among a small group of customers or consumers. Although popular articles can be a good recommendation, they are also likely to be well known and are sometimes even bad recommendations, especially in the face of pharmaceutical cross-selling recommendations. Furthermore, delivering only popular articles will not enhance the discovery of newly introduced articles and will ignore good recommendations that are contained in the long tail. The idea of the long tail of article popularity and its impact on recommendation quality has been explored by some researchers [21,29–32]. In those works, the authors tried to improve the performance of the recommender system in terms of accuracy and precision, while others focused on reducing popularity bias by means of regularization [33] or re-ranking [9]. Substantial research has also been published on recommendation diversity, where the goal is to prevent recommending too many similar articles [30–32].

2. Methodology

2.1. Data Representation in a Graph

In order to harness the power of graph convolutional neural networks to learn product embeddings, we transform the pharmacy sales data and product information into a graph. We represent each unique pharmacy product as a node, which also contains the descriptive information of the corresponding product encoded into a multi-hot vector, which has about 15,000 entries, representing the medical indications, active ingredients, and adverse reactions of the products. The data were extracted from the main German commercial pharmaceutical database used in pharmaceutical and other software solutions for pharmacies, physicians, hospitals, and other health providers, which was provided by Pharmatechnik. Undirected weighted edges between two nodes then represent how often cross-selling occurred for each pair of products, where a "cross-selling" is defined as two products sold in the same transaction. The set of approximately 100 million transactions with information about which products were sold together was provided by Pharmatechnik, leveraging sales transactions that are documented via their pharmacy management system IXOS, which is currently being used in more than 5000 pharmacies. Before further processing, we limit the transactions on which we base our cross-selling numbers to those with two or three sold articles, which is roughly 25% of the complete set. We hypothesize that the fewer items are in one transaction, the more specifically related the sold products are to each other. These transactions together include about 700,000 different products, but also include many similar products that vary only slightly (or not at all from each) other. For example, there are multiple offerings of Aspirin 100 products from different manufacturers. For training, validation, and testing, we then randomly chose 60% of these preselected transactions to construct a training graph and 20% each to construct validation and test graphs. These graphs act as the input to the model training and validation stage. However, the final model that is used for inference is then trained on all the included sales transactions. In model training, we use the top ranked cross-selling articles (those with the highest edge weights with a given query) as candidates for the aggregation and optimization part of the modeling process. See the details below. Popular articles unfortunately dominate this pool of candidates. To counterbalance this so-called popularity bias, we introduce a probability theory based approach that aims at updating the edge weights of all edges in the graphs.

2.1.1. Probability Based Re-Ranking

In the probability based re-ranking approach (PBR), we aim to discern the residual cross-sales from the expected cross-sales. This is achieved by subtracting the expected cross-sales of one product relative to another from their actual raw cross-sales. The residual cross-sales is now the new edge weight for the directed edge between both products.

Given products A and B, we aim to compute the conditional expected cross-sales of B given A, assuming both are independent. We approximate the conditional expected cross-sales of B given A, $E(B \mid A)$ to be:

$$E(B \mid A) = P(B, A \mid \mathcal{N}(a)) \cdot Z \tag{1}$$

The joint probability of A and B is conditioned on $\mathcal{N}(a)$, the subset of nodes in the one hop neighborhood of A. Conditioning on $\mathcal{N}(a)$ quantifies how often B is sold with A relative to how often the neighbors of A are sold with A. Under the assumption of the independence of A and B, $P(B, A \mid \cdot)$ in (1) can be reformulated as:

$$P(B, A \mid \mathcal{N}(a)) = P(B \mid \mathcal{N}(a)) \cdot P(A \mid \mathcal{N}(a)), \tag{2}$$

where we compute $P(B \mid \cdot)$, $P(A \mid \cdot)$, and Z as:

$$P(B \mid \mathcal{N}(a)) = \frac{\displaystyle\sum_{l \in \mathcal{N}(b)} \phi_{b,l}(b^1, l^1)}{Z}, \tag{3}$$

$$P(A \mid \mathcal{N}(a)) = \frac{\displaystyle\sum_{u \in \mathcal{N}(a)} \phi_{a,u}(a^1, u^1)}{Z}, \tag{4}$$

$$Z = \sum_{u \in \mathcal{N}(a)} \sum_{k \in \mathcal{N}(u)} \phi_{u,k}(u^1, k^1) \tag{5}$$

$\phi_{x,y}(x^1, y^1)$ is the factor representing the combination where x and y are sold together, and its value is the actual cross-selling amount between nodes x and y. To compute the conditional expected cross-sales, we first compute the cross-selling probability of A, $P(A \mid \cdot)$, as the total cross-sales amount of A divided by Z, which is the total amount of cross-sales in the one hop neighborhood of A; that is, the total number of cross-sales of the neighbors of A, which includes all cross-sales of A's neighbors with A. The cross-selling probability of B conditioned on the neighborhood of A, $P(B \mid \cdot)$, is computed in a similar manner, by dividing the total cross-sales of B by the same factor, Z. We then compute the conditional expected cross-sales (1) of B given A by multiplying the conditioned expected cross-selling probability of the two products $P(B \mid \cdot)$ and $P(A \mid \cdot)$, their joint probability in the case of independence (3) , with Z. To compute the residual cross-selling amount of B with respect to A, the conditional expected cross-selling is deducted from the actual cross-selling of A and B (4). This difference is the new edge weight between B and A.

$$W_{residual}(B, A) = W_{raw}(B, A) - E(B \mid A) \tag{6}$$

$W_{residual}$ is the final residual cross-selling amount of B with A, and $W_{raw}(B, A)$ is the cross-selling amount of nodes B and A from the input data. If $W_{residual}$ is positive, we assume that B is sold with A more often than expected; if $W_{residual}$ is negative, we assume that B is sold less often with A than expected. Note that $W_{residual}(A, B)$ is computed differently from $W_{residual}(B, A)$, as $P(B \mid \cdot)$ and $P(A \mid \cdot)$ are both conditioned on the one hop neighborhood of A. In the new graph, all nodes that were connected via undirected edges beforehand, therefore, now share two directed edges, expressing the residual cross-selling amount of both nodes relative to each other.

2.2. Model Architecture

We employ a graph convolutional neural network model (Figure 1) that uses localized convolutions on aggregated neighborhood vectors to produce embeddings of products represented by graph nodes, akin to the one introduced in [24]. The basic idea is that we transform the representations of the neighbors of a given node through a dense neural network and then apply an aggregator/pooling function (a weighted sum, shown as dark blue boxes in Figure 1, the "CONVOLVE" module) on the resulting set of vectors. This aggregation step provides a vector representation of a node's local neighborhood. We then concatenate this aggregated neighborhood vector (dark grey box) with the nodes' current representation (light grey box) and transform the concatenated vector through another dense neural network layer. The output of the algorithm is a representation of a node, called the node embedding, that incorporates information about itself and its local graph neighborhood. Details about the algorithm can be found in [22] (Algorithm 1); the only change we made is that we aggregate the neighborhood-node information only across the "top neighbors" of a given node. The top neighbor nodes of a given query node are those with the highest edge weight among all connected nodes, representing the products that are sold more often with the query than expected. We then compute the recommendations by using these final node embeddings, as described in Section 2.4.

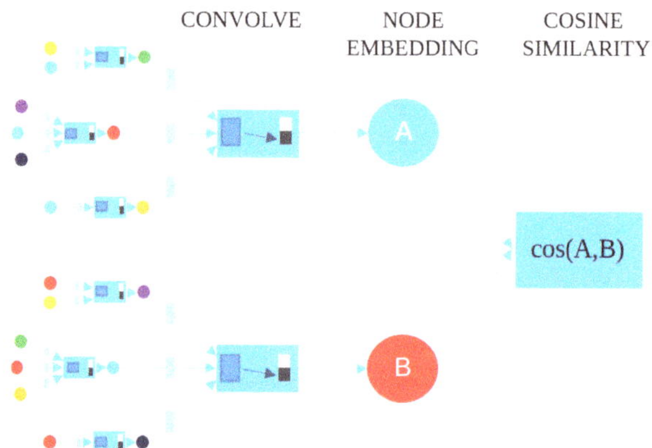

Figure 1. The graph convolutional neural network uses localized convolutions on aggregated neighborhood vectors to learn product embeddings. Here, we show the two layer graph neural network that computes the final embeddings of nodes A and B using the previous layer representation of nodes A and B, respectively, and that of their respective neighborhoods. Different colors denote different nodes. The recommendation score between two products A, B is then computed via the utilization of the cosine similarity between the two final embedding vectors of nodes A and B.

2.3. Model Training

For the optimization of the network parameters, we utilize the triplet loss, shown in Equation (6), which is a distance based loss function that operates on the final embedding of three input nodes:

the anchor A, the positive P, which is typically from the same class as the anchor or related by some other measure, and the negative N, which is typically from a different class than the anchor.

$$\min_\theta L(A, P, N) \tag{7}$$

$$L(A, P, N) = \max\left(0, sim(A, N) - sim(A, P) + \Delta\right) \tag{8}$$

We use the embeddings of pharmacy products represented by nodes in the graph as inputs for the triplet loss. Each node can be an anchor node A, and each has positively related nodes P and negatively related nodes N. A given pair A, P is to be related by some measure, i.e., are often bought together or are similar in feature encoding. The pair A, N is thought to be related by some other measure, i.e., are never bought together or are bought together less than expected. The goal of the training phase is to optimize the parameters of PharmaSage so that the cosine similarity $sim(A, P)$ of the anchor-positive pair is higher relative to the cosine similarity $sim(A, N)$ of the anchor-negative pair by a margin of 0.5.

2.3.1. Positive Sampling

We chose positives for a given anchor node A by randomly sampling nodes among the top neighbors of that anchor node with equal probability in each training iteration. The top ranked cross-sellers are those neighbors with the highest edge weight with the anchor, representing those products that have been sold more often with the anchor than expected. Together, they make up the positive sampling pool. To complement the edge-weight based positive sampling approach, we additionally utilize nodes that share encoding features with the anchor as possible positive samples. In this hybrid approach to positive sampling, we chose 50% of the positive samples based on the feature similarity of the anchor node relative to all other nodes and the other 50% based on edge weights representing re-ranked cross-sales. We only include products in the additional feature based positive sampling pool for a given anchor, if that product/node shares any feature with the anchor.

2.3.2. Negative Sampling

We chose negative samples among nodes not connected to the anchor node, which are never sold together with the anchor. We additionally utilized nodes with negative edge weights with the anchor as possible negative samples, as their expected cross-selling amount with the anchor is higher than the actual cross-selling. Generally, we applied semi-hard negative mining among the pool of possible negatives, as introduced in FaceNet [34], which has been widely used ever since [35,36].

2.4. Recommendation

The final embeddings that are computed by our model after training are then used to calculate recommendation scores between all products. We obtained a recommendation score bounded within the interval [0, 1] with the utilization of the cosine similarity of two vectors, $sim(A, B)$, commonly used in information retrieval and data mining techniques [37,38]. Given the similarity scores between one query product and all other products (except prescription medications, which can only be sold if prescribed by a physician, and hence are not legitimate cross-selling recommendations), the products with the highest similarity to the query are chosen as recommendations for the query article. A diagram illustrating the high-level architecture of how recommendations are computed between articles A and B is shown in Figure 1.

2.5. System Setup, Runtime, and Validation

For our software, we used the Python framework, the networkx library for creating the graph, and the PyTorch framework for the implementation of the GCN. The network is trained on a system with an Intel Core i7 8750H, an Nvidia GeForce GTX1070 GPU, and 32GB RAM. The training takes approximately 48 h and is terminated once the triplet loss, which starts at the margin of 0.5, has reached a threshold of 0.05 for the training graph. The validation loss is at 0.092 and the test loss at 0.095 at this point. We found that, empirically, the quality of the results does not improve much beyond this point. We stopped the training of the model used for inference at the same threshold.

3. Experiments

3.1. Popularity Bias in Sales Data

Popularity bias is a phenomenon that is visible in many data sources, including retail and online sales data. Our analysis of pharmacy sales data shows the same bias, as 2% of all products are sold equally often as the remaining 98%. It is clear in Figure 2a that those products that have a high overall sales probability also are ranked among the top neighbors on average, while unpopular articles have a proportionally lower cross-selling rank. This indicates that top-sellers are products that are also top cross-sellers.

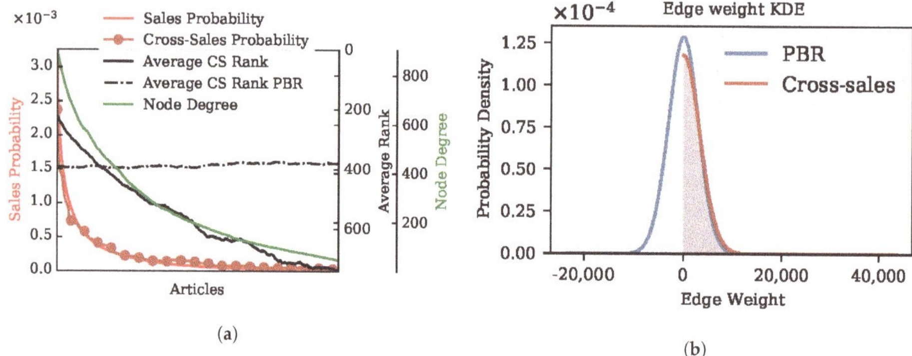

Figure 2. Popularity bias and re-ranking. (**a**) Depiction of overall sales probability and cross-sales (CS) probability ($y1$-axis) and the node degree (number of neighbors) ($y3$-axis) for all products. The average cross-sales rank ($y2$-axis) is shown for initial cross-sales statistics (solid black line) and the probability based re-ranking (PBR) approach (black dotted line). Note that the average rank is proportional to the popularity of the respective product in the initial cross-sales statistics, but this relation disappears in the PBR approach. (**b**) The probability density of edge weights (kernel density estimation (KDE)) shows the distribution of edge weights in the initial cross-sales based graph, as well as in the PBR based graph.

3.1.1. The Effect of Re-Ranking on Cross-Selling Statistics

Re-ranking affects the ranking of cross-sellers. As shown in Figure 2a, the average cross-sales rank for all products was influenced by re-ranking. The analysis of the graph based on actual cross-sales shows that the average rank of products decreases proportionally to the overall sales probability of that product. In contrast, the overall rank of a product has no correlation to its overall sales probability in the PBR approach. The average rank of top sellers drops substantially based on the PBR, which is a result of two factors. First, top cross-sellers are being sold often with many other products, as reflected by the number of neighbors' curve (node degree). Second, it is likely that they are only partly cross-sold more often than average. The same seems to be true for all other products and is reflected in the edge weight probability density after re-ranking (Figure 2b). After applying the PBR approach, the average weight across all nodes drops to 0 with a standard deviation of 308, indicating that half of the pairs are sold less often together than expected, while the other half represents genuine cross-sales. To further examine the effect of re-ranking on the distribution of ranks, Figure 3a shows two articles before and after re-ranking. Based on the actual cross-sales statistics, the median rank of the example top-seller is 7, but drops to 152 based on the PBR, respectively. The interquartile range based on the penalty re-ranking approach is 17.5 times larger than the PBR approach, which reflects a more dispersed ranking scheme. In contrast, the median rank of an example from the long tail is 562 and rises to 245 based on the PBR, with its interquartile range decreasing slightly.

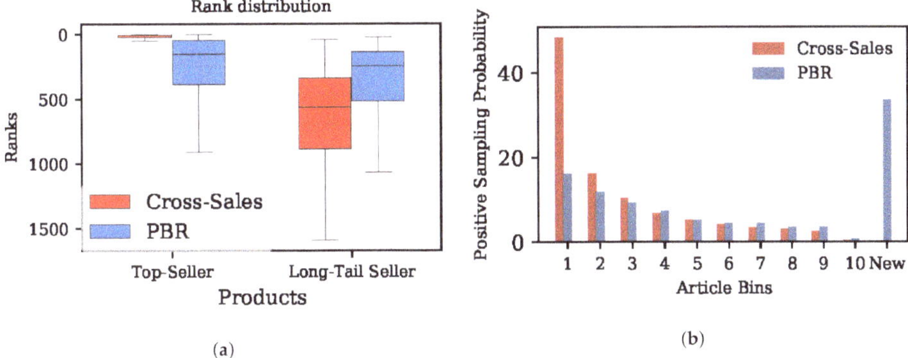

Figure 3. (**a**) Cross-selling rank distribution for a top-selling product and a product from the long tail based on raw cross-sales and the PBR. (**b**) Average probability for products (grouped into 10 quantiles) to be among the top cross-selling articles that are being used for positive sampling in the triplet loss for both approaches. "New" denotes products that are not included in positive sampling in the approach based on raw cross-sales, but are relevant in the PBR approach, and shows their average positive sampling probability. These products show up in addition among top ranked cross-selling products when the PBR is applied. The PBR introduces around 75% more products with an average sampling probability 36.6%. Quantile selection of products for the first 10 product bins for both approaches is based on raw cross-selling statistics (red).

3.1.2. The Effect of Re-Ranking on Positive Sampling

In order to assess the impact of the PBR on positive sampling involved in the network optimization, we evaluate how the positive sampling pool that is made up of the top neighbors of each node changes with the different approaches. The probability of being chosen among positive samples, as depicted in Figure 3b, shows that those products that dominate the positive sampling pool based on the raw cross-sales are used as positive samples significantly less often in the PBR approach. Sampling based on actual cross-sales statistics puts more emphasis on popular products, while the PBR puts less emphasis on those products, but adds more diversity to the sampling pool. Relative to the raw cross-selling statistics, the PBR introduces more diversity, with 36.6% additional products in the sampling pool.

3.2. Recommendation Quality

In order to evaluate the performance of our algorithm, we collected expert feedback on product-product recommendations for a set of 25 query products. For each recommendation generated by PharmaSage, the feedback given can have one of three values; 0 (no pharmaceutical relation to the query product), 1 (a pharmaceutical relation exists, and the product is a good recommendation for the query), or 2 (a pharmaceutical relation exists, and the product is a very good recommendation for the query). The mean across scores for the top 15 recommendations for 25 products is then used as a performance quantifier.

First, we compute recommendations based on the cross-selling data encoded in the graph, which can be thought of as a conventional recommendation approach akin to collaborative filtering. Products that are most often sold with the query product are chosen as recommendations. Second, we compute the recommendations based on the PBR graph. This result offers insight into what can be achieved without any additional learning. We then use the PBR graph to train the PharmaSage model.

As can be seen in Figure 4, the quality of the evaluated recommendations is comparatively lower for simple cross-sales statistics based recommendations, and the quality increases in the PBR approach. The PharmaSage model that is then trained on the PBR approach introduces yet another increase in recommendation quality. Compared to approaches based solely on sales data, this model is able to learn from cross-sale information encoded in the graph edges, as well as leverage feature information encoded as node features.

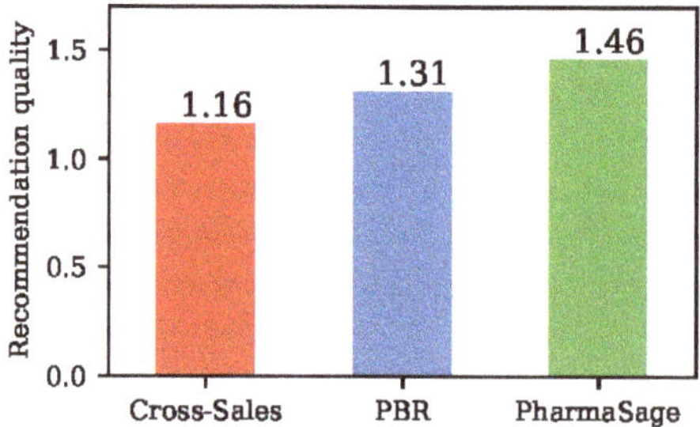

Figure 4. Average recommendation quality among the top 15 recommended articles for 25 evaluated products. Recommendations are computed based on the graph incorporating raw cross-selling statistics and the graph, where the edge weights have been recomputed using the probability approach (PBR). PharmaSage is optimized based on the PBR approach as the input.

In Table 1, example recommendations computed by PharmaSage, their recommendation rank, and expert feedback are shown for prednisolone, a corticosteroid. Prednisolone is used to treat a wide range of health problems including allergies, skin diseases, infections, and certain autoimmune disorders. It helps by reducing inflammation and suppressing the immune system. PharmaSage recommends additional over-the-counter medications for allergies like hay fever (rank 1, 5, Table 1) for which Prednisolone is often prescribed and stomach acid blockers to reduce the adverse effect of the active ingredient in the query article (rank 2, 4, Table 1). Calcium and vitamin D3 can both help improve bone health, which can be affected by continuous therapy with corticosteroids (rank 6, 9, Table 1). The nonsteroidal anti-inflammatory drug helps with pain and induces additional anti-inflammatory effects (rank 8, Table 1). Furthermore, an antipruritic and anti-inflammatory gel (rank 10, Table 1) is recommended, which counters symptoms associated with the main indications of prednisolone, and a nasal spray is also for additional therapy, which frees up a clogged nose, reduces the swelling of the mucous membranes, and can help treat allergic reactions of the respiratory tract (rank 3, Table 1). ASS100 (rank 7, Table 1), an anticoagulant, should not be recommended without additional medical advice; hence, the expert feedback mark is 0.

Table 1. Example recommendations for prednisolone, a corticosteroid.

Rank	Recommended Product	Expert Feedback
1	CETIRIZIN AL DIREKT	2
2	PANTOPRAZOL ABZ	2
3	OTRIVEN 0.1	2
4	OMEPRADEX 20MG	2
5	LORANO AKUT	2
6	DEKRISTOL 400 IE	2
7	ASS AL 100 TAH	0
8	IBUPROFEN OPT 400MG	2
9	CALCIUM D3 RATIO	2
10	FENISTIL	2

4. Conclusions

We introduced PharmaSage, a graph convolutional network for pharmacy product cross-sale recommendations. PharmaSage is the first application of graph convolutional neural networks to pharmaceutical product-product recommendation, exploiting sales statistics and pharmaceutical product features like indications, ingredients, and adverse effects. In addition, we introduced a method based on probability theory that addresses the common popularity bias problem. We showed how popularity bias is present in the pharmacies' sales and cross-selling dataset and how it can be successfully addressed in order to increase both recommendation quality and diversity. We developed PharmaSage based on real-world pharmaceutical data and comprehensively evaluated the quality of the learned embeddings for a pharmacy product cross-sales recommendation task, demonstrating a substantial improvement in recommendation quality compared to traditional approaches that are based solely on cross-sales statistics. Our work demonstrates the positive impact that methods based on graph convolutional networks can have in pharmacy cross-selling recommender systems, and we believe that PharmaSage can be further extended to tackle other graph representation learning problems in the retail and online sales industry. A future point of interest will be to further evaluate the quality of recommendations given by PharmaSage by evaluating A/B tests against traditional recommendations given by pharmacists and how both impact cross-sales and evaluating customer feedback.

Author Contributions: Methodology, software, and validation: F.H., Y.T., G.H., and H.M.; data curation: S.H.; review, editing, and supervision: F.H., G.H., H.M., S.H. and A.K.; All authors read and agreed to the published version of the manuscript.

Funding: This research received no external funding.

Conflicts of Interest: The authors declare no conflict of interest.

References

1. Hamilton, W.L.; Ying, R.; Leskovec, J. Representation learning on graphs: Methods and applications. *arXiv* **2017**, arXiv:1709.05584.
2. Kipf, T.N.; Welling, M. Semi-supervised classification with graph convolutional networks. *arXiv* **2016**, arXiv:1609.02907.
3. Monti, F.; Bronstein, M.; Bresson, X. Geometric matrix completion with recurrent multi-graph neural networks. In Proceedings of the Neural Information Processing Systems, Long Beach, CA, USA, 4–9 December 2017; pp. 3697–3707.
4. van den Berg, R.; Kipf, T.N.; Welling, M. Graph convolutional matrix completion. In Proceedings of the KDD'18 Deep Learning Day, London, UK, 19–23 August 2018.
5. Covington, P.; Adams, J.; Sargin, E. Deep neural networks for youtube recommendations. In Proceedings of the 10th ACM Conference on Recommender Systems, Boston, MA, USA, 15–19 September 2016; pp. 191–198.
6. Van den Oord, A.; Dieleman, S.; Schrauwen, B. Deep content based music recommendation. In Proceedings of the Neural Information Processing Systems, Lake Tahoe, NV, USA, 5–10 December 2013; pp. 2643–2651.
7. Zitnik, M.; Agrawal, M.; Leskovec, J. Modeling polypharmacy side effects with graph convolutional networks. *Bioinformatics* **2018**, *34*, i457–i466. [CrossRef] [PubMed]
8. Kearnes, S.; McCloskey, K.; Berndl, M.; Pande, V.; Riley, P. Molecular graph convolutions: Moving beyond fingerprints. *J. Comput. Aided Mol. Des.* **2016**, *30*, 595–608. [CrossRef] [PubMed]
9. Himan Abdollahpouri, R.B.; Bamshad, M. Managing popularity bias in recommender systems with personalized re-ranking. In Proceedings of the Thirty-Second International Flairs Conference, Sarasota, FL, USA, 19–22 May 2019.
10. Paul Rutter, E.W. Does evidence drive pharmacist over-the-counter product recommendations? *J. Eval. Clin. Pract.* **2013**, *20*, 1365–2753.
11. Sarah, E.C.; Hans, W. Factors affecting pharmacists' recommendation of complementary medicines—A qualitative pilot study of Australian pharmacists. *BMC Complement. Altern. Med.* **2012**, *12*, 183.

12. Clayton, K.; Luxford, Y.; Colaci, J.; Hasan, M.; Miltiadou, R.; Novikova, D.; Vlahopoulos, D.; Stupans, I. Community pharmacists' recommendations for natural products for stress in Melbourne, Australia: A simulated patient study. *Pharm Pract.* **2020**, *18*, 1660. [CrossRef]

13. Bhat, S.; Aishwarya, K. Item based Hybrid Recommender System for newly marketed pharmaceutical drugs. In Proceedings of the International Conference on Advances in Computing, Communications and Informatics, Kerala, India, 29–31 August 2013; pp. 2107–2111.

14. Zhang, Y.; Zhang, D.; Hassan, M.M.; Alamri, A.; Peng, L. CADRE: Cloud-Assisted Drug Recommendation Service for Online Pharmacies. *Mob. Netw. Appl.* **2014**, *20*, 348–355. [CrossRef]

15. Dai, H.; Dai, B.; Song, L. Discriminative embeddings of latent variable models for structured data. In Proceedings of the International Conference on Machine Learning, New York, NY, USA, 19–24 June 2016; pp. 2702–2711.

16. Xu, K.; Hu, W.; Leskovec, J.; Jegelka, S. How powerful are graph neural networks? In Proceedings of the International Conference on Learning Representations, New Orleans, LA, USA, 6–9 May 2019.

17. Bruna, J.; Zaremba, W.; Szlam, A.; LeCun, Y. Spectral networks and locally connected networks on graphs. In Proceedings of the International Conference on Learning Representations, Banff, AB, Canada, 14–16 April 2014

18. Bronstein, M.M.; Bruna, J.; LeCun, Y.; Szlam, A.; Vandergheynst, P.T. Geometric deep learning: Going beyond euclidean data. *IEEE Signal Process. Mag.* **2017**, *34*, 18–42. [CrossRef]

19. Defferrard, M.; Bresson, X.; Vandergheynst, P. Convolutional neural networks on graphs with fast localized spectral filtering. In Proceedings of the Neural Information Processing Systems, Barcelona, Spain, 5–10 December 2016; pp. 3844–3852.

20. Duvenaud, D.K.; Maclaurin, D.; Iparraguirre, J.; Bombarell, R.; Hirzel, T.; Aspuru-Guzik, A.; Adams, R.P. Convolutional networks on graphs for learning molecular fingerprints. In Proceedings of the Neural Information Processing Systems, Montreal, QC, Canada, 7–12 December 2015; pp. 2224–2232.

21. You, J.; Ying, R.; Ren, X.; Hamilton, W.L.; Leskovec, J. Graphrnn: Generating realistic graphs with deep auto-regressive models. In Proceedings of the 35th International Conference on Machine Learning, Stockholm, Sweden, 10–15 July 2018.

22. Ying, R.; He, R.; Chen, K.; Eksombatchai, P.; Hamilton, W.L.; Leskovec, J. Graph convolutional neural networks for web-scale recommender systems. In Proceedings of the 24th ACM SIGKDD International Conference on Knowledge Discovery & Data Mining, London, UK, 19–23 August 2018; pp. 974–983.

23. Perozzi, B.; Al-Rfou, R.; Skiena, S. Deepwalk: Online learning of social representations. In Proceedings of the 20th ACM SIGKDD International Conference on Knowledge Discovery and Data Mining, New York, NY, USA, 24–27 August 2014; pp. 701–710.

24. Hamilton, W.; Ying, Z.; Leskovec, J. Inductive representation learning on large graphs. In Proceedings of the Neural Information Processing Systems, Long Beach, CA, USA, 4–9 December 2017; pp. 1024–1034.

25. Veličković, P.; Cucurull, G.; Casanova, A.; Romero, A.; Lio, P.; Bengio, Y. Graph attention networks. In Proceedings of the International Conference on Learning Representations, Vancouver, BC, Canada, 30 April–3 May 2018.

26. Seongjun, Y.; Minbyu, J.; Raehyun, K.; Jaewoo, K.; Hyunwoo, J.K. Graph Transformer Networks. In Proceedings of the 33rd Conference on Neural Information Processing Systems, Vancouver, BC, Canada, 8–14 December 2019.

27. Bellogín, A.; Castells, P.; Cantador, I. Statistical biases in Information Retrieval metrics for recommender systems. *Inf. Retr. J.* **2017**, *20*, 606–634. [CrossRef]

28. Park, Y.-J.; Tuzhilin, A. The long tail of recommender systems and how to leverage it. In Proceedings of the 2008 ACM Conference on Recommender Systems, Lausanne, Switzerland, 23–25 October 2008; pp. 11–18.

29. Brynjolfsson, E.; Hu, Y.J.; Smith, M.D. From niches to riches: Anatomy of the long tail. *IEEE Signal Process. Mag.* **2006**, *47*, 67–71.

30. Zhou, T.; Kuscsik, Z.; Liu, J.-G.; Medo, M.; Wakeling, J.R.; Zhang, Y.-C. Solving the apparent diversity-accuracy dilemma of recommender systems. *Proc. Natl. Acad. Sci. USA* **2010**, *104*, 511–4515. [CrossRef] [PubMed]

31. Castells, P.; Vargas, S.; Wang, J. Rank and Relevance in Novelty and Diversity Metrics for Recommender Systems. In Proceedings of the ACM Conference on Recommender Systems, Chicago, IL, USA, 23–27 October 2011.

32. Zhang, M.; Hurley, N. Avoiding monotony: Improving the diversity of recommendation lists. In Proceedings of the 2008 ACM Conference on Recommender Systems, Lausanne, Switzerland, 23–25 October 2008; pp. 123–130.

33. Abdollahpouri, H.; Burke, R.; Mobasher, B. Controlling popularity bias in learning-to-rank recommendation. In Proceedings of the Eleventh ACM Conference on Recommender Systems, Como, Italy, 27–31 August 2017; pp. 42–46.

34. Schroff, F.; Kalenichenko, D.; Philbin, J. Facenet: A unified embedding for face recognition and clustering. In Proceedings of the IEEE Conference on Computer Vision and Pattern Recognition, Boston, MA, USA, 7–12 June 2015; pp. 815–823.

35. Oh Song, H.; Xiang, Y.; Jegelka, S.; Savarese, S. Deep metric learning via lifted structured feature embedding. In Proceedings of the IEEE Conference on Computer Vision and Pattern Recognition, Las Vegas, NV, USA, 27–30 June 2016; pp. 4004–4012.

36. Parkhi, O.M.; Vedaldi, A.; Zisserman, A. Deep face recognition. In Proceedings of the British Machine Vision Conference (BMVC), Swansea, UK, 7–10 September 2015; pp. 41.1–41.12.

37. Singhal, A. Modern information retrieval: A brief overview. *IEEE Data Eng. Bull.* **2017**, *24*, 35–43.

38. Tan, P.-N.; Steinbach, M.; Kumar, V. Data mining cluster analysis: Basic concepts and algorithms. In *Introduction to Data Mining*; University of Minnesota: Minneapolis, MN, USA, 2013; pp. 487–533.

Publisher's Note: MDPI stays neutral with regard to jurisdictional claims in published maps and institutional affiliations.

Article

Exploiting the User Social Context to Address Neighborhood Bias in Collaborative Filtering Music Recommender Systems

Diego Sánchez-Moreno, Vivian López Batista, M. Dolores Muñoz Vicente,
Ángel Luis Sánchez Lázaro and María N. Moreno-García *

Department of Computer Science and Automation, University of Salamanca, 37008 Salamanca, Spain;
sanchez91@usal.es (D.S.-M.); vivian@usal.es (V.L.B.); mariado@usal.es (M.D.M.V.); alsl@usal.es (Á.L.S.L.)
* Correspondence: mmg@usal.es

Received: 11 August 2020; Accepted: 10 September 2020; Published: 11 September 2020

Abstract: Recent research in the field of recommender systems focuses on the incorporation of social information into collaborative filtering methods to improve the reliability of recommendations. Social networks enclose valuable data regarding user behavior and connections that can be exploited in this area to infer knowledge about user preferences and social influence. The fact that streaming music platforms have some social functionalities also allows this type of information to be used for music recommendation. In this work, we take advantage of the friendship structure to address a type of recommendation bias derived from the way collaborative filtering methods compute the neighborhood. These methods restrict the rating predictions for a user to the items that have been rated by their nearest neighbors while leaving out other items that might be of his/her interest. This problem is different from the popularity bias caused by the power-law distribution of the item rating frequency (long-tail), well-known in the music domain, although both shortcomings can be related. Our proposal is based on extending and diversifying the neighborhood by capturing trust and homophily effects between users through social structure metrics. The results show an increase in potentially recommendable items while reducing recommendation error rates.

Keywords: music recommender systems; social influence; social trust; homophily; collaborative filtering; streaming services

---

1. Introduction

Social networks are currently the focus of intensive research, as they are a great source of information that can be used in multiple domains for multiple purposes. Recommender systems are one of the areas in which social data can be exploited to improve the reliability of recommendations. The adoption of streaming music services as a common way of listening to music has allowed its use in this domain since most of these platforms are, in turn, equipped with some kind of social functionality, such as establishing friendship connections. In addition, streaming systems collect user interactions, which allows implicit feedback from users to be used instead of explicit ratings as an expression of user preferences. This has promoted the development of recommender systems for these platforms. Nevertheless, the implementation of methods that take advantage of social information is scarce in the music streaming services environment, because the mechanisms of social interaction are much more limited than in social networks such as Facebook, Twitter, etc.

Currently, the methods most extensively used in recommender systems are based on Collaborative Filtering (CF). This approach requires either explicit or implicit user ratings or preferences for some products that users have already consumed. The larger is the number of ratings, the higher is the reliability of the recommendations provided by these methods. Many proposals that make use of social data are precisely aimed at minimizing the drawback of insufficient ratings, while others are just focused on improving rating prediction without dealing with problems concerning recommendation bias.

In this work, we introduce the concept of neighborhood bias that takes place in the context of collaborative filtering methods and causes a limitation in the number of potentially recommendable items. In these approaches, the recommendations made to a given user are restricted to items rated by other users with similar tastes, who are called his/her nearest neighbors. This fact prevents the user from discovering other items that he/she might like. The neighborhood bias is caused by the way neighbors are found since this process is based on the similarity of users' ratings about the same products. For example, two users may have the same musical tastes, but those users cannot be neighbors if they have rated different artists or songs. This problem is related to popularity bias because it is more likely that the most popular items are the most rated and, therefore, the most recommended. However, the bias that we try to address in this work is not the same since the objective is to extend the range of potentially recommended items but not necessarily with the less popular items. To achieve this, we propose to extend the neighborhood by considering social factors that may have some impact on user preferences. Thus, the neighborhood of a given user is calculated not only on the basis of affinity in preferences with other users, but also on the influence received from other users in the social network. When extending the neighborhood using social factors, the number of potentially recommendable items is also extended, since the greater is the number of neighbors, the greater is the number of items with which they interact. Trust and homophily are two factors that influence users when choosing products or services, and, therefore, must be considered when predicting their interests and preferences. Trust refers to individuals who are more likely to adopt recommendations not only from opinion leaders but from their closest social context, while homophily refers to the similarity of connected users in social networks since they usually share tastes and interests. The graph of social connections between users can be the subject of structural measures that capture these two factors and allow their influence on recommendations to be considered. Many methods have been proposed in the literature for such purpose, although it has been shown that their performance depends largely on the application domain [1] and most of them have been validated in specific domains other than music [2,3]. In the music area, they have not been sufficiently tested, mainly due to the difficulty of obtaining the necessary social information from streaming platforms. For instance, friendship connections are only bidirectional, and there is only one between each pair of users. This makes it impossible to apply well known graph-based metrics, such as centrality, page-rank, etc., which work with unidirectional connections, to establish two connections between each pair of users, one in each direction. In addition, that information has usually been used to improve the reliability of recommendations. It has not been exploited to deal with neighborhood bias, which is the main purpose of our work. It is therefore necessary to develop effective techniques to obtain these factors in this environment to integrate them into traditional recommendation methods and benefit from them.

This work addresses the problem of incorporating social information obtained from music streaming systems into CF methods to improve the recommendations provided to users. Although their reliability is taken into account, the improvement is mainly focused on widening their variety by dealing with the problem of the neighborhood bias, which has great importance in this type of systems. To achieve this objective, social structure metrics that capture the concepts of trust and homophily are incorporated into the recommendation process. Our approach differs from existing ones in that social information is not used to modify the value of rating predictions but to complement them. In addition, this proposal significantly improves predictions, while most proposals focus on extending the variety of recommendations at the cost of losing accuracy or maintaining it at best. This improvement is also achieved by using only the limited social data available on streaming platforms.

Another aspect addressed in this paper is the lack of explicit ratings on music items. This inconvenience is overcome by calculating implicit ratings from the frequency of plays, recorded by streaming systems.

The rest of the paper is organized as follows. Section 2 presents a summary of the related work. The approach to incorporate structural metrics into CF methods is described in Section 3.

The experimental study conducted to validate the proposal and the discussion of results are included in Section 4. Finally, the conclusions are presented in Section 5.

2. Related Work

Both the use of social information and dealing with bias is the focus of much recent research in recommender systems, but there is little work in the literature in which both topics are addressed together. The objective generally pursued in studies that exploit social information is to improve recommendations by including social data processing into the rating prediction method so that the predicted value is closer to the actual value. This is mostly done by modifying either neighborhood-based CF techniques [4,5] or matrix factorization methods [6]. Regarding the work facing recommendation bias, the main proposals involve data preprocessing as resampling or clustering or postprocessing procedures as reranking, as set out below.

Bias in machine learning models is a widely studied and discussed problem that can be seen from different perspectives. Several types of bias have been studied in the recommender systems area, although most are related to unfair recommendations, from race or gender discrimination [7] to popularity bias [8]. In the former, the problem is usually addressed through recommendation algorithms that are sensitive to this bias and focus on the protection of discriminated groups. Burke et al. [9] introduced the concept of a balanced neighborhood with respect to the protected and unprotected classes to enhance the fairness of recommendations without compromising personalization. In our work context, some artists in the music domain may be harmed by biased recommendations, while user satisfaction may be affected by the limited choice of items that can be recommended to them, especially to the so-called grey sheep users whose tastes are unusual. However, these unfair recommendations are not associated with any specific attribute, such as gender or race.

Popularity bias is mainly associated with neighborhood-based methods, the most frequently used, and is one of the major concerns of recent research in this field. There are proposals for facing this problem that focus on improving recommendations for grey sheep users [10,11], while others are focused on increasing the recommendations of the less frequently rated items and improving item diversity. This can be achieved through probabilistic models [12], data preprocessing [13] or postprocessing [14,15]. There are studies that address aggregate diversity that refers not only to diversity of individual recommendations but also across recommendations of all users [16,17]. The aim of these studies is to improve diversity while maintaining accuracy or with a minimum loss of it. Our proposal is different since it is a user-centered approach, which aims to expand the possibilities of the items to be recommended, but, in this case, by diversifying the user neighborhood. This is done by drawing on factors, such as trust and homophily, derived from the social network structure.

The concept of social trust is the most studied in the literature about recommender systems. It is usually used to give more relevance to the ratings of trusted users against others [18] since it can be considered as a form of social influence that is often obtained from friendship connections, comments, messages, etc. Some systems allow users to explicitly express their trust on opinions, reviews and comments given by other users, but, in most cases, this is not possible, and it is necessary to infer it implicitly [1].

Social trust can be used locally when only opinions of connected friends are taken into account and globally when reputed individuals in the entire network are considered [19]. On the other hand, some approaches use social trust without considering similarity between users, while, in others, it is used jointly with similarity values [20,21] or even with additional factors, such as different types of interactions in social networks [19]. There are many works in the literature where diverse factors affecting social influence are addressed, but most of them are focused on social networks such as Facebook or Twitter, from which a great variety of social information can be extracted.

Homophily and trust are two related concepts [22]. The effect of homophily can even be used for trust prediction [23], although homophily effects have been less studied and are often included in the general study of social influence without explicitly differentiating. Some recent work analyzes the influence of homophily on consumers' purchasing decisions in the context of YouTube and Instagram influencers' popularity [24,25]. However, in these works, homophily is treated as a complex factor that encompasses aspects such as attitude, background, morality and appearance. Therefore, it cannot only be inferred from the structure of social relations. In the area of recommender systems, the study of homophily is much scarcer. In [3], recommendations of tourist attractions are generated by classifying users into several types, depending on factors such as homophily. This factor is determined by the membership of users in social communities.

Although trust and homophily principles have been much less studied in the field of music recommendation, we can highlight the work of Fields et al. [26], where music recommendations are based on the social relevance of musical data obtained through complex network technologies. A different objective is pursued in [23], in which the factors influencing the music listening homophily are analyzed. The analysis includes social information and user demographic attributes. None of these studies have addressed the problem of bias in the recommendations.

This section describes relevant work that is closely related to the proposal presented here. However, current approaches to improving recommender systems are many and varied. Among them is the promising field of cognitive computing that would allow an interaction between users and recommender systems similar to human interaction [27]. Emotion and sentiment analysis is also being widely used in the recommendation area, especially in context-aware systems where recommendations depend on the emotional state of the user [28]. Although social information can be used to infer emotions, it is usually textual information from comments or reviews that is not always available [29]. Another trend in this field, although more distant from our proposal, is the research on binary codes that is focused on efficiency and storage optimization in large-scale recommender systems [30].

3. Incorporating Social Structure Metrics into User-Based Collaborative Filtering

Collaborative filtering methods is to predict how much a user would like an item from the ratings that other users have given to that item. User-based or user-user collaborative filtering methods base the recommendations on the similarity between users, considering that two users are similar if they have similarly rated the same items.

Given a set of m users $U = \{u_1, u_2, \ldots, u_m\}$ and a set of n items $I = \{i_1, i_2, \ldots, i_n\}$, each user u_i has a list of ratings that he/she has given to a set of items I_{ui}, where $I_{ui} \subseteq I$. In this context, a recommendation for the active user $u_a \in U$ involves a set of items I_{ui}, where $I_{ui} \subseteq I$. In this context, a recommendation for the active user $u_a \in U$ involves a set of items $I_{ra} \subset I$ that fulfill the condition $I_{ra} \cap I_{ua} = \emptyset$, since only items not rated by u_a can be recommended. The similarity between users is computed from ratings by means of different distance-based measures such as cosine, Chebyshev and Jaccard or correlations coefficients such as Pearson, Kendall and Spearman. Among them, the most extensively used in the field of recommender systems are the Pearson coefficient and cosine similarity. The similarity between the active user u_a and another user u_i is denoted as $sim(u_a, u_i)$.

$$sim(u_a, u_i)_{Pearson} = \frac{\sum_j (r_{aj} - \bar{r}_a)(r_{ij} - \bar{r}_i)}{\sqrt{\sum_j (r_{aj} - \bar{r}_a)^2 (r_{ij} - \bar{r}_i)^2}} \tag{1}$$

where r_{aj} and r_{ij} are the ratings of user u_a and user u_i for item i_j, respectively, and $\bar{r}_a$ and $\bar{r}_i$ are the average ratings of user u_a and user u_i, respectively. The Pearson coefficient can represent inverse and direct correlation with its values in the interval $[-1, 1]$, where the value 0 corresponds to the absence of correlation.

The well-known cosine similarity metric for two given users, u_a and u_i, is computed according to Equation (2), where $V_{u_a} R_{u_a}$ and V_{u_i} are the vectors containing the ratings given to items by users u_a and u_i, respectively.

$$sim(u_a, u_i) = cos\left(V_{u_a}, V_{u_i}\right) = \frac{V_{u_a} \cdot V_{u_i}}{\|V_{u_a}\| \|V_{u_i}\|} \tag{2}$$

The items recommended to the active user are the best evaluated by the users most similar to him/her.

CF methods can be improved by introducing social information. Trust and homophily are two factors influencing the recommendations that can be inferred from the structure of relationships between users and other social network resources. However, in most music streaming services, those resources are much more limited, and the structure is restricted to bidirectional friendship relations, which does not allow centrality, page-rank and other graph-based metrics to be applied. In this work, we use the friendship structure to derive trust and homophily factors to include them in the recommendation process.

3.1. Structural Equivalence for Homophily Inference

Structural equivalence is a property applicable to social communities in social networks, often used to identify implicit communities by computing the equivalence similarity between pairs of nodes in the network. Equivalence similarity is based on the overlap between the neighborhood of those nodes. In the context of this work, this metric can be applied to friendship structure whose nodes represent the users. Nodes with high similarity are considered to be part of the same implicit community. This is a way to capture the homophily concept since users belonging to the same community usually share interests and preferences. Therefore, their ratings can be used by the recommendation methods.

Let us consider two nodes representing two users u_i and u_j of the social network, and $N(u_i)$ and $N(u_j)$ their respective neighborhoods. In this context, two users are only considered neighbors if there is a direct link between them in the friendship structure. A measure of the similarity between a pair of nodes can be defined in terms of the neighbors common to both, as follow:

$$socialSim_{abs}(u_i, u_j) = \left| N(u_i) \cap N(u_j) \right| \tag{3}$$

To get a similarity value in the range [0, 1], some metrics, such as Jaccard or cosine, can be used for normalization (Equations (4) and (5)). These similarities are used together with the similarities derived from the ratings in the framework proposed in this work.

$$socialSim_{Jaccard}(u_i, u_j) = \frac{\left| N(u_i) \cap N(u_j) \right|}{\left| N(u_i) \cup N(u_j) \right|} \tag{4}$$

$$socialSim_{Cosine}(u_i, u_j) = \frac{\left| N(u_i) \cap N(u_j) \right|}{\sqrt{\left| N(u_i) \right| \left| N(u_j) \right|}} \tag{5}$$

A possible problem with the structural equivalency measure lies in the fact that nodes u_i and u_j are excluded from both neighborhoods. Therefore, if those nodes are directly connected and their similarity is very low or even zero, those nodes would not belong to the same community. This is not a drawback in our study since direct friend relationships are also treated in the proposed recommendation approach. The way to approach these types of connections is explained below.

3.2. Friendship Connections for Trust Inference

There are some systems in which users are allowed to make revisions about products and other users can explicitly express their confidence in them by rating such revisions. However, these mechanisms are not available in most systems, so trust has to be inferred from comments, relationships and other types of interaction between users.

On streaming music platforms, bidirectional friendship relationships can be used to infer trust. In the same way that people ask their friends for opinions in the real-world and are influenced by them, users are influenced locally by other users through the friendship connections they establish in social networks. It can be said that users have more trust in those users directly linked to them than the rest. Social trust can be used to improve recommender systems. However, due to the trust that friends exert, the influence is not the same in all circumstances but depends on many factors. If we only focus on the social structure, we can infer trust from the friendship connections.

Any user of the streaming platform can be connected directly with other users of the platform who we call friends. The set of friends of a user $u_i \in U$ is denoted as $F_i \subseteq U$.

We are assuming that the trust of one user in another depends on the influence that the latter has on the former. On the one hand, it seems reasonable that the influence of friends on a given user is greater the fewer friends he has. On the other hand, those users who are more influential are those who have more friends. Taking these premises into account, we can obtain a function that represents the degree of trust that a user has in another user belonging to his group of friends. To establish the relationship between influence and number of friends, we define for each user u_i a logarithmic function of the number of friends:

$$logF(u_i) = \log(|F_i|) \tag{6}$$

where F_i is the set of friends of the user u_i.

From the above equation, we can define the trust of the active user u_a in any of the users $u_f \in F_a$ connected to him directly, that is, his friends in the social network.

$$t\left(u_a, u_f\right) = \frac{1}{1 + logF(u_a)} logF\left(u_f\right) \tag{7}$$

These values are also used in the proposed recommendation process, which is presented in Section 3.3.

3.3. Recommendation Method Based on Social Structure Metrics

In most recommendation methods that exploit the user's social context, social information is used to modify the value of the predictions for a given item: it can be used by modifying the similarity between users based on ratings, as a function that combines predictions based on ratings and those based on social information is applied, as social regularization term added to the rating-based function used to make predictions, etc.

The approach proposed in this paper is very different, since our purpose is to use social structure metrics that capture trust and homophily to complement predictions based on ratings, in order to increase the number and variety of recommended items while also increasing the reliability of the recommendations.

The proposed algorithm combines three types of recommendations: based on rating similarity, based on social equivalence similarity and based on friend influence.

3.3.1. Recommendations Based on Rating Similarity

These types of recommendations are those made in traditional CF systems. The procedure for obtaining them is detailed below.

Let us consider the set U of m users and the set I, of n items. Each user $u_i \in U$ has rated or interacted with a subset of items $I_{ui} \subseteq I$. Ratings are stored in an $m \times n$ matrix $\mathbf{R}$ called the rating matrix, where each element is the rating that a user u_i gives to an item j.

$$\mathbf{R} := r_{i,j}, \ \mathbf{R} \in M_{m \times n}(\mathbb{N}) \tag{8}$$

When explicit ratings are not available or they are scarce, some strategies to compute implicit ratings can be used. In the field of music, where the items to be recommended are artists or songs, a common way is to calculate them from the frequency of plays. In our case, instead of using binary or simple frequency functions, we apply a linear function of the frequency percentile [31]. In this method, the play frequency for a given user u_i and an item (artist/song) i_j is computed from an $m \times n$ matrix of plays $:= p_{i,j}$, which is analogous to the rating matrix, but contains the number of plays of each user for each artist/song. The play frequency is defined as follows:

$$pFreq_{i,j} = \frac{p_{i,j}}{\sum_{j'} p_{i,j'}} \qquad (9)$$

where $p_{i,j}$ is the number of times that a user u_i plays an artist/song i_j and j' represents each of the items (artists/songs) played by user u_i.

These items are ordered by their frequency values for the user u_i. $Freq_{k'}(i)$ denotes the frequency $Freq_{i,j}$ of an item i_j with rank k, being $k' = 1$ for the artist/song having the highest frequency. A rating for an item with rank k is computed as a linear function of the frequency percentile:

$$r_{i,j} = 4\left(1 - \sum_{k=1}^{k-1} pFreq_k(i)\right) \qquad (10)$$

The factor with value 4 in the equation is used to obtain rating values in the interval $(0, 4]$.

These implicit ratings are used in the same way as the explicit ones in the CF methods. When using this approach, the recommendations for the active user u_a are calculated from the ratings of other users by means of techniques such as k Nearest Neighbors (k-NN). They require computing the similarity between users by using some of the available metrics. The most used, Pearson correlation coefficient and cosine similarity can be computed by means of Equations (1) and (2), respectively.

The similarity between the active users and the rest of the users, $sim(u_a, u_i)$, calculated by any of the metrics, is used to predict the rating that the active user would give to an item i_j that he/she has not played yet, by means of Equation (11). Only the set $k\_NN_a \subseteq U$ of k nearest neighbors, that is, those with the highest similarity values, will be taken to make the predictions pr_{aj}.

$$pr_{aj} = \bar{r}_a + \frac{\sum_{i=1}^{k} sim(u_a, u_i)\left(r_{ij} - \bar{r}_i\right)}{\sum_{i=1}^{k}\left|sim(u_a, u_i)\right|} \qquad (11)$$

where $\bar{r}_a$ and $\bar{r}_i$ are average values for user u_a and user u_i, respectively:

$$\bar{r}_i = \frac{1}{|I_i|} \sum_{j \in I_i} r_{ij} \qquad (12)$$

The recommendations obtained by this method are those used as a starting point in the approach proposed in this paper.

3.3.2. Recommendations Based on Structural Equivalence

A problem that occurs when making the rating prediction for a certain item j by applying the previous procedure derives from the fact that some items whose ratings for the active we want to predict, have not been evaluated by the nearest neighbors. This introduces the neighborhood bias that greatly limits the number of potentially recommendable items for a given user. To address this drawback, we make use of the measures related to structural equivalence that can be obtained from the friendship network.

We introduce the concept of social similarity, $sim_{social}(u_i, u_j)$, as the similarity obtained with any of the equivalence metrics defined in Section 3.1 (Equations (4) and (5)). Social similarity is used jointly with similarity based on ratings $sim_{rat}(u_i, u_j)$, using some combination function. The combined similarity $sim_c(u_i, u_j)$ is used to find a different set $k\_NN_{sociala}$ of k nearest neighbors for the active user u_a, which is defined as follows:

$$k\_NN_{sociala} \subseteq U \qquad (13)$$

$$k\_NN_{sociala} = \{u_1, u_2, \ldots, u_k\} \tag{14}$$

where $sim_c(u_a, u_1) > sim_c(u_a, u_2) > \ldots > sim_c(u_a, u_k)$. We use Equation (15) to calculate the combined similarity. In this way, we use the homophily concept to find new users in the social environment of the active user with potentially similar preferences, and whose ratings can be used in the recommendation process.

$$sim_c(u_i, u_j) = \sqrt{\alpha\, sim_{rat}(u_i, u_j)^2 + (1 - \alpha)\, sim_{social}(u_i, u_j)^2} \tag{15}$$

To make the predictions pr_{aj} in this case, Equation (11) is also used, but utilizing the combined similarity and with a different set of neighbors. The set $k\_NN_{sociala}$ obtained from social similarities is used instead of the set based on rating similarity.

3.3.3. Recommendations Based on Friendship Connections

Analogous to the process described in the previous subsection, we can exploit the concept of trust derived from the friendship connections to find users who are likely to influence the preferences of the active user. Then, a new subset of k nearest neighbors, $k\_NN_{friends}$, is formed with the most influential friends of the active user.

To determine the degree of influence or trust, $t(u_a, u_f)$, of a friend $u_f \in F_a$ on the active user u_a, we make use of the Equations (6) and (7). This value is used in Equation (18) to predict the ratings for the active user. Within the set of friends of u_a, the subset of the k nearest neighbors used to compute the predictions for u_a is defined as follows:

$$k\_NN_{friendsa} \subseteq U \tag{16}$$

$$k\_NN_{friendsa} = \{u_1, u_2, \ldots, u_k\} \mid t(u_a, u_1) \rangle\, t(u_a, u_2) > \ldots > t(u_a, u_k) \tag{17}$$

This type of recommendations, based on trust between friends, are obtained by using the set of nearest neighbors $k\_NN_{friendsa}$ and Equation (18), in which the rating-based similarity is multiplied by a weight given by that trust.

$$pr_{aj} = \bar{r}_a + \frac{\sum_{i=1}^{k} t(u_a, u_f)\, sim(u_a, u_i)(r_{ij} - \bar{r}_i)}{\sum_{i=1}^{k} \left| t(u_a, u_f) sim(u_a, u_i) \right|} \tag{18}$$

3.3.4. Recommendation Algorithm

The idea of the approach proposed in this paper is to complement the recommendations generated by traditional collaborative filtering methods with recommendations based on the structure of the users' social network. Thus, all types of recommendations described in the previous subsections are involved in the recommendation algorithm. The goal is to increase the number of predictions in order to expand the set of potentially recommendable items and reduce the neighborhood bias, while improving the reliability of the recommendations. The algorithm is shown in Figure 1.

```
1:  Input:  P := p_{i,j},  P ∈ M_{n×m}(ℕ)| u_i ∈ U, i_j ∈ I  // Matrix of plays
2:          F := f_{i,j},  P ∈ M_{m×m}(ℕ)| u_i, u_j ∈ U  // Matrix of friends
3:  // Computing implicit ratings
4:  compute Freq_{i,j}
5:  for i = 1 to n  do
6:          T = sequence { (Freq_{i,j})_k } ∀ j | p_{i,j} > 0 ∧ (Freq_{i,j})_k > (Freq_{i,j})_{k+1}
7:          S(i) = sequence {Freq_k(i)} = T
8:              for j = 1 to m  do
9:                  Set k value | Freq_k(i) = Freq_{i,j}
10:                 r_{i,j} = 4 (1 − Σ_{k'=1}^{k-1} Freq_k(i))
11:             end for
12: end for
13: set R := r_{i,j} , R ∈ M_{n×m}(ℕ)
14: compute similarity matrix S_r := sim_{rat}(u_i, u_j) , S_r ∈ M_{m×m}(ℕ), u_i, u_j ∈ U
15: compute social similarity matrix S_s := sim_{social}(u_i, u_j) , S_s ∈ M_{m×m}(ℕ), u_i, u_j ∈ U
16: compute combined similarity matrix S_c := sim_c(u_i, u_j) S_c ∈ M_{m×m}(ℕ), u_i, u_j ∈ U
17: compute t(u_i, u_j)
18: set k
19: for i = 1 to m  do
20:         compute k_NN_i
21:         compute k_NN_{social_i}
22:       compute k_NN_{friends_i}
23: end for
24: //Recommendations for user u_a
25: Set u_a ∈ U
26: compute Ī_a = I − I_r: = i_{a,j}  // set of items not rated by u_a
27: for j = 1 to |Ī_a|  do
28:         if ∃ N ∈ k_NN_a | j ∈ I_N
29:             compute pr_{aj} using k_NN_a and eq. 11 with sim_{rat}(u_i, u_j)
30:         else
31:             if ∃ N ∈ k_NN_{friends_a} | j ∈ I_N
32:               Compute pr_{aj} using k_NN_{friends_a} and eq. 18
33:             else
34:                 if ∃ N ∈ k_NN_{social_a} | j ∈ I_N
35:                     compute pr_{aj} using k_NN_{social_a} and eq. 11 with sim_c(u_i, u_j)
36:             end if
37: end for
38: Output: Pr_a: = pr_{a,j}
```

Figure 1. Algorithm for the complete recommendation process, including the calculation of implicit ratings.

The only input data required by the algorithm is the matrix of plays $\mathbf{P} := p_{i,j}$, defined in Section 3.3.1, and the matrix of friends $\mathbf{F}$, defined as follows:

$$\mathbf{F} := f_{i,j}, \mathbf{F} \in M_{m \times m}(\mathbb{N}) \, \middle| \, f_{i,j} = \begin{cases} 1, & u_i \text{ is friend of } u_j \\ 0, & \text{otherwise} \end{cases} \tag{19}$$

Prior to the recommendation process, it is necessary to calculate the implicit ratings from the play matrix according to the procedure described in Section 3.3.1. Steps 3–13 of the algorithm are those corresponding to this calculation from which the matrix of ratings $\mathbf{R}$ is obtained.

Similarities based on ratings, social and combined similarities and trust $t(u_i, u_j)$ between users are calculated in Steps 14–17. Steps 17–23 are devoted to obtaining the different sets of k nearest neighbors. First, the value of k is set, and then the sets k_NN_i, $k\_NN_{sociali}$ and $k\_NN_{friendsi}$ for each user u_i are created.

Subsequent steps contain the complete recommendation process for a given active user u_a. The basic CF method is first applied using the set of the nearest neighbors k_NN_a obtained from the rating-based similarities, according to the procedure described in Section 3.3.1. The number of predictions pr_{aj} obtained in this way is lower than all the possible ones since many items have not been rated by the users who are in the set k_NN_a. To achieve a greater number of predictions, the procedures defined for making both predictions based on structural equivalence (Section 3.3.2) and predictions based on friendship connections (Section 3.3.3) are applied. This last one is applied first to the items without predicted ratings by the basic CF method, and the ratings for the remainder items are tried to be predicted from the set of k nearest neighbor $k\_NN_{sociala}$.

4. Validation of the Proposed Approach

4.1. Dataset

Since our proposal is specifically designed for the field of music recommendation, its validation was carried out with a dataset obtained from Hetrec2011-lastfm [32]. The only information needed to apply the recommendation method is the data about the playing songs by users, in particular the number of plays, as well as the friendship connections between users in the social network of the streaming system.

The play frequency is used to compute implicit ratings according to the procedure described in Section 3.3.1. The availability of implicit or explicit ratings is a prerequisite for applying CF techniques since user similarities are based on these ratings. Friendship connections are required to compute social structure metrics, used in this work to extend the basic CF methods in the previously explained manner.

4.2. Baseline Methods

To validate the proposed method, its results were compared with those of other proposals in the literature. For this purpose, two methods that do not use social information and two other methods that make use of information inferred from the friendship structure of the social network were tested with the same dataset. Among the former, the most representative ones were chosen, user-based k-NN and matrix factorization. Among those that exploit social information, the baseline methods were an approach in which CF is constrained to the user social context (SCC) and another that combines social similarities and rating-based similarities (SSW).

The tested user-based k-NN method is the same one described in Section 3.3.1, while matrix factorization is a well-known technique in the area of recommender systems.

The methods that constrain CF to the social context are those in which the set of nearest neighbors is formed only with users connected directly to the active user u_a, i.e., their friends ($k\_NN \subseteq F_a$) [22]. Similarity metrics used to find the neighbors are based on ratings, no social similarity metrics are applied. The procedure for making predictions can be the same as in user-based k-NN (Equation (11)).

Regarding the last type of methods, these make use of some function to combine social similarities, $sim_{social}(u_i, u_j)$, *and* rating-based similarities, $sim_{rat}(u_i, u_j)$. Then, the final similarity is used in the prediction of ratings. In this case, user-based k-NN and Equation (11) can also be utilized. The set k_NN is created by using the combined similarity $sim_c(u_i, u_j)$ defined by Equation (15).

The specific social similarity used in our study is based on structural equivalence and α was set to 0.7.

There are other methods that extend matrix factorization approaches to incorporate social data, but we only tested k-NN-based approaches because these give better results than matrix factorization, as can be seen in the following section.

4.3. Empirical Study

This study was conducted to compare the proposed approach against the baseline methods. The metrics used to evaluate rating prediction reliability were Root-Mean-Square Error (RMSE), Mean Absolute Error (MAE), Normalized RMSE (NRMSE) and Normalized MAE (NMAE). Mean Average Precision (MAP) and Normalized Discounted Cumulative Gain (NDCG) were used for the evaluation of top-N recommendations. In all the experiments, five-fold cross validation was applied.

The first step of the study was to determine the number of nearest neighbors to use in k-NN-based methods. Thus, the results of the application of the user-based k-NN method were compared with a variable number of neighbors, from 10 to 40. Figure 2 shows the error rates produced. Since the increase of k value from 20 produces a very small decrease in errors, we decided to conduct the tests with k = 20.

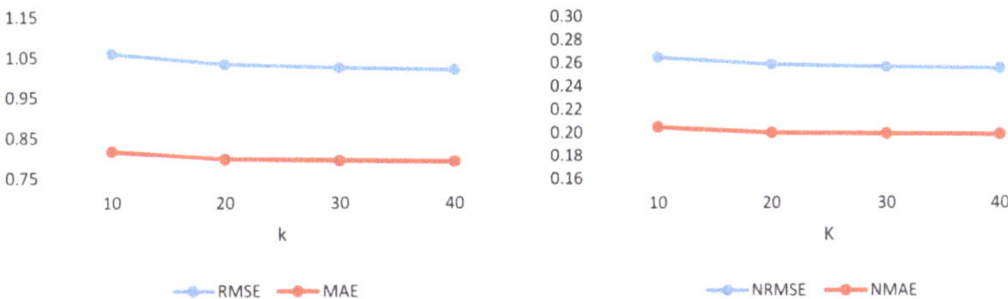

Figure 2. Error rates of user-based k-NN for different values of k.

The comparative study of different methods was then carried out to validate our proposal. The value k = 20 was used in all methods, except for Matrix Factorization (MF) since it is not a k-NN-based technique. In addition, the rating-based similarities of all k-NN-based methods were calculated using the cosine metric. Social similarities were obtained with the Jaccard equivalence similarity metric.

In addition to the study carried out with all the users in the dataset, we also studied the behavior of our proposal for the cold-start scenario. As mentioned above, this is a problem of CF methods that occurs mainly with new users because they have a few ratings/interactions with the items. In that case, the recommendations they receive are not very reliable.

To determine the performance on the cold-start scenario, the users with a low number of plays in relation to the other users were selected. Only the records of these users were kept in the test sets of all the folds, the rest was eliminated. In this case three-cross-validation was applied since the number of users was much lower. Taking into account that the average number of plays per user in the dataset is 37,275, the users with fewer than 2000 plays were selected.

Figure 3 shows the error rates of the methods. When comparing the two basic methods MF and user k-NN, a better behavior of the latter is observed. Regarding the methods that exploit social information, we see that no improvements are obtained with respect to k-NN but the results are even worse, both with CF restricted to the user social context (SCC) and with the method using weighted social and rating-based similarities (SSW). However, our proposal to combine user k-NN with Social Structure Metrics (user k-NN SSM) provides a clear improvement over all baseline methods. Table 1 shows this percentage improvement for RMSE and MAE.

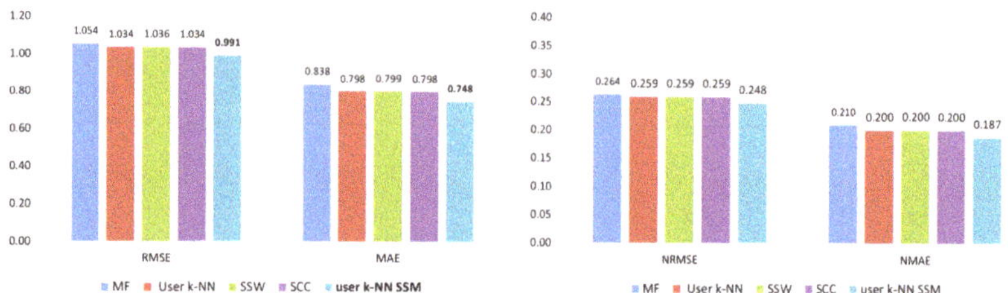

Figure 3. Error rates of the tested methods: Matrix Factorization (MF), user k-NN, CF with Social Similarity Weight (SSW), CF with Social Context Constraint (SCC) and the proposed approach user k-NN combined with Social Structure Metrics (user k-NN SSM).

Table 1. Improvement of user k-NN SSM over the baseline methods for all users (left) and for users with the cold-start problem (right).

	All Users		Cold-Start	
Baseline Method	**RMSE**	**MAE**	**RMSE**	**MAE**
MF	6.00%	10.79%	7.06%	13.96%
User k-NN	4.18%	6.32%	9.51%	14.76%
SSW	4.34%	6.46%	9.66%	14.77%
SCC	4.18%	6.32%	9.51%	14.76%

Figure 4 presents the results obtained in the cold-start scenario. As expected, the errors are higher than those obtained with all users, although user k-NN SSM is again the method with the lowest error rates. We can see in Table 1 that the percentages of improvement are even better for this scenario than for the previous one. The figure also shows that, in this case, MF provides better results than the other KNN-based methods. This can be explained by the fact that matrix factorization approaches behave better against sparsity.

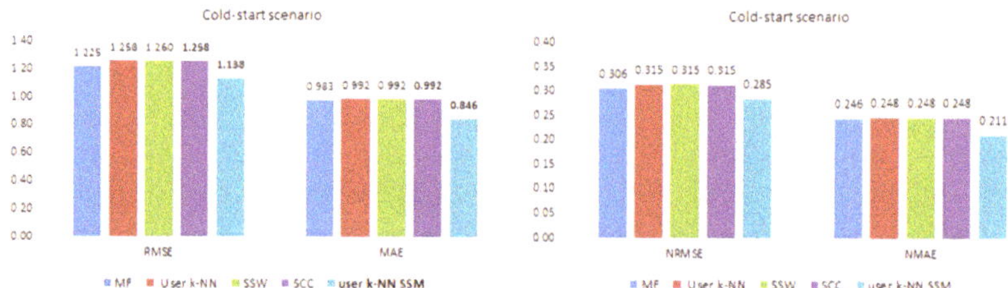

Figure 4. Error rates for the cold-start scenario.

However, the goal of our work is not only to improve the reliability of the rating predictions, but also to increase their number in order to have more potentially recommendable items. This way we would be able to increase the variety of recommendations and minimize the bias toward the most popular items.

A way to increase the number of predictions while decreasing their errors is to work with larger sets of nearest neighbors, although we previously showed that the improvement in predictions is very small above 20 neighbors. Figure 5 shows this decrease in error rates for user k-NN from 10 to 40 neighbors, as well as for the proposed method, user k-NN SSM, with 20 neighbors. We can see that the lowest error rates are given by our proposal even compared to user k-NN errors with a larger number of neighbors.

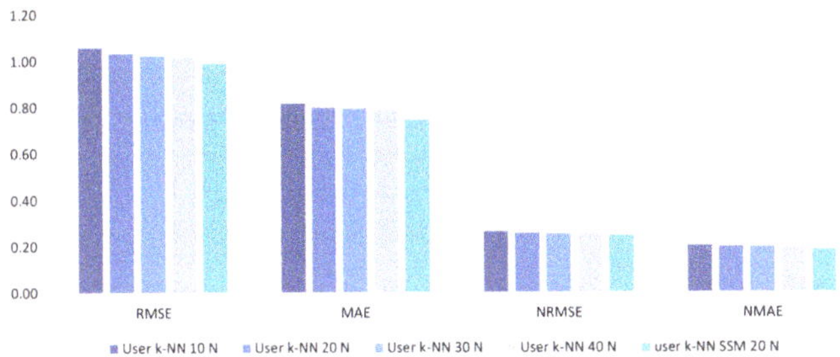

Figure 5. Error rates for user-based k-NN with a variable number of neighbors (10–40) and for user k-NN SS with 20 neighbors.

Since the main objective of the work is to increase the number of potentially recommendable items, we must demonstrate that our approach covers more rating predictions on the items in the test set. Figure 6 shows the coverage for both methods in each cross-validation fold. The graph on the left shows the results obtained for all users and the graph on the right those for users with few plays (cold-start). This graph clearly shows the significant increase in coverage over the k-NN method. Most methods that focus on expanding coverage result in increased error rates and their goal is usually to keep this increase to a minimum. In the case of our proposal, however, the errors actually decrease.

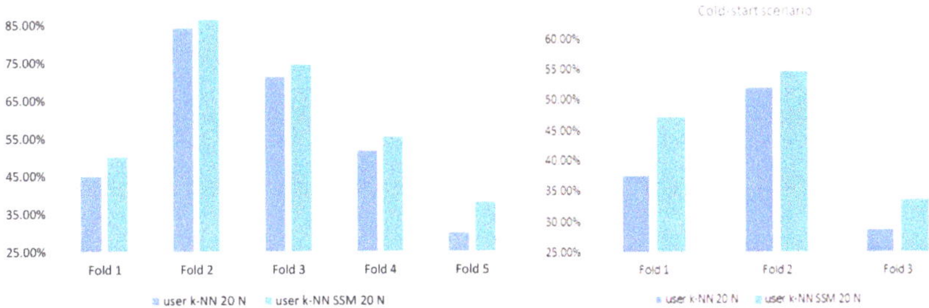

Figure 6. Coverage of the predictions for user-based k-NN and user k-NN SSM with 20 neighbors: (left) for all users; and (right) for the cold-start scenario.

Finally, to confirm the validity of the approach presented in this paper, the evaluation was also performed for top-N recommendations. In the rating prediction validation, errors were calculated for all predicted ratings. However, it is also necessary to make the validation for the lists of items with the highest ratings values because those items are the ones that are recommended to the user. Thus, we ensure that the higher reliability of the proposal is not only due to the predictions of low values but also to the predictions of the high values that are the most interesting for recommendation. We used the rank-based metrics MAP and NDCG for top-N lists where N was set to 5. Figure 7 shows these results, which prove that the best performance of the proposed method is also achieved for top-N recommendations. The behavior is similar in both scenarios, although, as in the case of rating prediction, in the cold-start scenario lower values of these metrics are obtained.

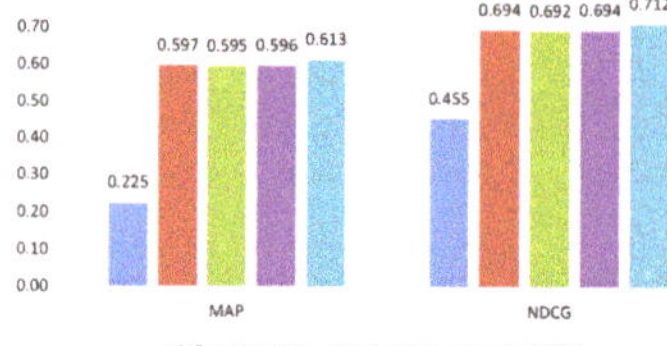
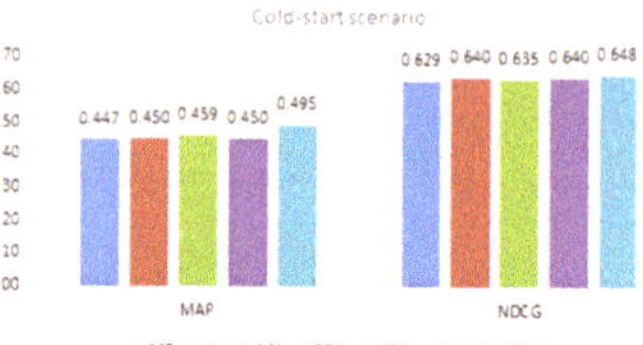

Figure 7. Evaluation of top-N recommendations by means of MAP and NDCG obtained with the baseline methods and the user k-NN SSM proposal for the two studied scenarios.

4.4. Discussion

The above results prove that the proposed method increases the coverage of the recommendations in relation to the potentially recommendable items. Furthermore, this increase does not come at the expense of the recommendation reliability but, on the contrary, results in an error decrease in predicting the ratings of these items as well as in an increase in the values of the rank-based metrics used to evaluate the quality of the recommended top-N lists. In the different proposals in the literature related to our work, as far as we know, both improvements are not obtained together. In addition, most of these works address the popularity bias while our aim is to increase the spectrum of potentially recommendable items regardless of whether these items are popular or not. We also did not find any work that uses social connections to expand the neighborhood in CF methods. Below, we discuss the differences between our proposal and some relevant work aimed at improving the diversity of the recommendations.

The re-ranking approach [14], which involves changing the ranking of items, addresses the popularity bias and improves recommendation diversity, but at the expense of recommendation accuracy. The local scoring model presented in [33] aiming at dealing with scalability and sparsity problems, provides a more efficient way to select the best neighbors and improves the recommendation diversity without compromising accuracy. In [16], a graph-based method that maximizes diversity for a given level of accuracy is presented. In the trade-off between diversity and accuracy shown in the work, it can be seen that as the diversity increases, the accuracy decreases, although more slowly than using the re-ranking-based methods. A more recent graph-based approach also focused on increase the recommendations of unpopular items is proposed in [17]. Although this proposal also does not improve accuracy when increasing diversity, it manages to maintain it, which is an important advantage over other methods. The calibration, a problem related to diversity is studied in [13]. The purpose is dealing with the problem that recommendations are biased to the main areas of interest of a user instead of proportionally reflecting the different interests of the user. This work shows that as the degree of calibration increases, the accuracy decreases.

Since the objective of the previous work is not exactly the same as ours, we cannot make comparisons with the coverage results obtained with our proposal. However, none of these methods improve accuracy and ours does. Another difference that could be considered a disadvantage of our method with respect to others is the need for social information, in addition to ratings. Although this information is restricted to friendship connections and is easily obtainable from streaming platforms.

5. Conclusions

The growing use of music streaming services and the interest in their personalization is unquestionable nowadays. Thus, this is one of the main motives why the surge in intensive research in many areas on the exploitation of information from social networks has been extended to music recommender systems.

In this work, an approach focused on exploiting social information available on streaming music platforms is proposed. It is a collaborative filtering scheme that extends classical methods based on nearest neighbors by using structural metrics obtained from the network of user friendships.

The goal is to minimize the neighborhood bias as well as to increase the reliability of recommendations. The proposal differs from others in the literature in the fact that it is a user-centered approach instead of being centered on items. In addition, it is not specifically addressed to increase the diversity or reduce popularity bias but to extend and diversify the user neighborhood by exploiting user social context. The results show that the proposed approach outperforms other methods in both reducing prediction error rates and increasing the number of potentially recommendable items.

Author Contributions: Conceptualization, D.S.-M., M.N.M.-G. and V.L.B.; methodology, D.S.-M., M.D.M.V. and Á.L.S.L.; software, D.S.-M.; validation, D.S.-M., M.D.M.V. and Á.L.S.L.; formal analysis, D.S.-M. and M.N.M.-G.; investigation, D.S.-M., M.N.M.-G., V.L.B., M.D.M.V. and Á.L.S.L.; data curation, D.S.-M.; writing—original draft preparation, D.S.-M.; writing—review and editing, M.N.M.-G.; supervision, M.N.M.-G.; project administration, M.N.M.-G.; and funding acquisition, M.N.M.-G. All authors have read and agreed to the published version of the manuscript.

Funding: This research was funded by Junta de Castilla y León, Spain, grant number SA064G19.

Conflicts of Interest: The authors declare no conflict of interest. The funders had no role in the design of the study; in the collection, analyses, or interpretation of data; in the writing of the manuscript, or in the decision to publish the results.

References

1. Aggarwal, C.C. *Recommender Systems*; Springer: New York, NY, USA, 2016.
2. Pérez-Marcos, J.; Martín-Gómez, L.; Jimenez-Bravo, D.M.; López, V.F.; Moreno-García, M.N. Hybrid system for video game recommendation based on implicit ratings and social networks. *J. Ambient Intell. Humaniz. Comput.* **2020**. [CrossRef]
3. Esmaeili, L.; Mardani, S.; Golpayegani, S.-A.H.; Madar, Z.Z. A novel tourism recommender system in the context of social commerce. *Expert Syst. Appl.* **2020**. [CrossRef]
4. Sánchez-Moreno, D.; Moreno-García, M.N.; Sonboli, N.; Mobasher, B.; Burke, R. Inferring user expertise from social tagging in music recommender systems for streaming services. In *Hybrid Artificial Intelligence Systems, Lecture Notes in Artificial Intelligence*; De Cos Juez, F.J., Villar, J.R., De la Cal, E.A., Herrero, A., Quintian, H., Saez, J.A., Corchado, E., Eds.; Springer: New York, NY, USA, 2018; pp. 39–49.
5. Sánchez-Moreno, D.; Pérez-Marcos, J.; Gil, A.B.; López, V.F.; Moreno-García, M.N. Social influence-based similarity measures for user-user collaborative filtering applied to music recommendation. In *Advances in Intelligent Systems and Computing: Distributed Computing and Artificial Intelligence, Special Sessions, 15th International, Conference, 2018*; Rodríguez, S., Prieto, J., Faria, P., Klos, S., Fernandez, A., Mazuelas, S., Jimenez-Lopez, M.D., Moreno, M.N., Navarro, E.M., Eds.; Springer: New York, NY, USA, 2019; pp. 1–8.
6. Yadav, P.; Rani, K.S.; Kumari, S. Review of social collaborative filtering recommender system's methods. *Intern. J. Eng. Comput. Sci.* **2015**, *4*, 14927–14932.
7. Mansoury, M.; Abdollahpouri, H.; Smith, J.; Dehpanah, A.; Pechenizkiy, M.; Mobasher, B. Investigating potential factors associated with gender discrimination in collaborative recommender systems. In Proceedings of the 13th FLAIRS Conference, Miami, FL, USA, 17–20 May 2020; pp. 193–196.
8. Abdollahpouri, H.; Mansoury, M.; Burke, R.; Mobasher, B. The unfairness of popularity bias in recommendation. In Proceedings of the 13th ACM conference on recommender systems (RecSys), 2019, RMSE Workshop, Copenhagen, Denmark, 20 September 2019.
9. Burke, R.; Sonboli, N.; Ordoñez-Gauger, A. Balanced neighborhoods for multi-sided fairness in recommendation, proceedings of machine learning research, proceedings of the 1st Conference on Fairness, Accountability, and Transparency. *Proc. Mach. Learn. Res.* **2018**, *81*, 1–13.
10. Sánchez-Moreno, D.; Gil, A.B.; Muñoz, M.D.; López, V.F.; Moreno, M.N. Recommendation of songs in music streaming services. Dealing with sparsity and gray sheep problems. In *Trends in Cyber-Physical Multi-Agent Systems. The PAAMS Collection–15th International Conference, PAAMS 2017. Advances in Intelligent Systems and Computing Series*; Springer International Publishing: Berlin/Heidelberg, Germany, 2017; Volume 619, pp. 206–213.
11. Sánchez-Moreno, D.; Muñoz, M.D.; López, V.F.; Gil, A.B.; Moreno, M.N. A session-based song recommendation approach involving user characterization along the play power-law distribution complexity. *Hindawi J.* **2020**. [CrossRef]

12. Vargas, S.; Castells, P. Rank and relevance in novelty and diversity metrics for recommender systems. In Proceedings of the 5th ACM Conference on Recommender Systems (RecSys), Chicago, IL, USA, 23–27 October 2011; pp. 109–116. [CrossRef]

13. Steck, H. Calibrated recommendations. In Proceedings of the 12th ACM Conference on Recommender Systems (RecSys), Vancouver, BC, Canada, 2–7 October 2018; pp. 154–162.

14. Adomavicius, G.; Kwon, Y. Toward more diverse recommendations: Item re-ranking methods for recommender systems. In Proceedings of the 19th Workshop on Information Technologies and Systems, Phoenix, AZ, USA, 14–15 December 2009.

15. Kamishima, T.; Akaho, S.; Asoh, H.; Sakuma, J. Recommendation independence. In Proceedings of the 1st Conference on Fairness, Accountability and Transparency (FAT), New York, NY, USA; 2018; pp. 187–201.

16. Adomavicius, G.; Kwon, Y. Maximizing aggregate recommendation diversity: A graph-theoretic approach. In Proceedings of the 1st International Workshop on Novelty and Diversity in Recommender Systems (DiveRS 2011), Chicago, IL, USA; 2011; pp. 3–10.

17. Mansoury, M.; Abdollahpouri, H.; Pechenizkiy, M.; Mobasher, B.; Burke, R. FairMatch: A graph-based approach for improving aggregate diversity in recommender systems. *arXiv* **2020**, arXiv:2005.01148.

18. Massa, P.; Avesani, P. Trust–aware recommender systems. In Proceedings of the ACM Conference on Recommender Systems RecSys, Minneapolis, MN, USA, 19–20 October 2007; pp. 17–24.

19. Kalaï, A.; Abdelghani, W.; Zayani, C.A.; Amous, I. LoTrust: A social trust level model based on time-aware social interactions and interests similarity. In Proceedings of the 14th IEEE Fourteenth Annual Conference on Privacy, Security and Trust, Auckland, New Zeland, 12–14 December 2016; pp. 428–436.

20. Akcora, C.G.; Carminati, B.; Ferrari, E. User similarities on social networks. *Soc. Netw. Anal. Min.* **2013**, *3*, 475–495. [CrossRef]

21. Ziegler, C.; Golbeck, J. Investigating interactions of trust and interest similarity. *Decis. Support Syst.* **2006**, *43*, 460–475. [CrossRef]

22. Zafarani, R.; Abbasi, M.A.; Liu, H. *Social Media Mining*; Cambridge University Press: Cambridge, UK, 2014.

23. Zhou, Z.; Xu, K.; Zhao, J. Homophily of music listening in online social networks of China. *Soc. Netw.* **2018**, *55*, 160–169. [CrossRef]

24. Ladhari, R.; Massa, E.; Skandrani, H. YouTube vloggers' popularity and influence: The roles of homophily, emotional attachment, and expertise. *J. Retail. Consum. Serv.* **2020**, *54*. [CrossRef]

25. Sokolova, K.; Kefi, H. Instagram and YouTube bloggers promote it, why should I buy? How credibility and parasocial interaction influence purchase intentions. *J. Retail. Consum. Serv.* **2020**, *53*. [CrossRef]

26. Fields, B.; Jacobson, K.; Rhodes, C.; Inverno, M.; Sanler, M.; Casey, M. Analysis and exploitation of musician social networks for recommendation and discovery. *IEEE Trans. Multimed.* **2011**, *13*, 674–686. [CrossRef]

27. Angulo, C.; Falomir, I.Z.; Anguita, D.; Agell, N. Bridging cognitive models and recommender systems. *Cogn. Comput.* **2020**, *12*, 426–427. [CrossRef]

28. Zheng, Y.; Mobasher, B.; Burke, R. Emotions in context-aware recommender systems. In *Emotions and Personality in Personalized Services*; Tkalčič, M., De Carolis, B., de Gemmis, M., Kosir, A., Odic, A., Eds.; Springer: New York, NY, USA, 2016.

29. Hung, B.T. Integrating sentiment analysis in recommender systems. In *Reliability and Statistical Computing: Springer Series in Reliability Engineering*; Pham, H., Ed.; Springer: Cham, Switzerland, 2020.

30. Zhang, Y.; Wu, J.; Wang, H. Neural binary representation learning for large-scale collaborative filtering. *IEEE Access* **2019**, *7*, 60–752. [CrossRef]

31. Pacula, M. A Matrix Factorization Algorithm for Music Recommendation Using Implicit User Feedback. Available online: http://www.mpacula.com/publications/lastfm.pdf. (accessed on 11 September 2020).

32. Cantador, I.; Brusilovsky, P.; Kuflik, T. 2nd Hetrec workshop. In Proceedings of the 5th ACM Conference on Recommender Systems, RecSys, New York, NY, USA, 23–27 October 2011.

33. Kim, H.K.; Kim, J.K.; Ryu, Y. A local scoring model for recommendation. In Proceedings of the 20th Workshop on Information Technologies and Systems (WITS'10), Paphos, Cyprus, 27–28 March 2010.

Article

Detecting and Tracking Significant Events for Individuals on Twitter by Monitoring the Evolution of Twitter Followership Networks

Tao Tang [1] and Guangmin Hu [2],*

[1] School of Information and Communication Engineering, University of Electronic Science and Technology of China, Chengdu 611731, China; terotongcn@gmail.com

[2] School of Resources and Environment, University of Electronic Science and Technology of China, Chengdu 611731, China

* Correspondence: hgm@uestc.edu.cn; Tel.: +86-028-6183-0209

Received: 24 August 2020; Accepted: 15 September 2020; Published: 16 September 2020

Abstract: People publish tweets on Twitter to share everything from global news to their daily life. Abundant user-generated content makes Twitter become one of the major channels for people to obtain information about real-world events. Event detection techniques help to extract events from massive amounts of Twitter data. However, most existing techniques are based on Twitter information streams, which contain plenty of noise and polluted content that would affect the accuracy of the detecting result. In this article, we present an event discovery method based on the change of the user's followers, which can detect the occurrences of significant events relevant to the particular user. We divide these events into categories according to the positive or negative effect on the specific user. Further, we observe the evolution of individuals' followership networks and analyze the dynamics of networks. The results show that events have different effects on the evolution of different features of Twitter followership networks. Our findings may play an important role for realizing how patterns of social interaction are impacted by events and can be applied in fields such as public opinion monitoring, disaster warning, crisis management, and intelligent decision making.

Keywords: ego network; events; network dynamics; Twitter

1. Introduction

Twitter is one of the most popular online social media platforms with more than 300 million monthly active users. Users are allowed to publish short messages no more than 140 characters called tweets to express what is happening in the world and their opinions about it. A user can build directed social connections to others representing the "follow" relationship. The initiator of a connection is called the follower, and the recipient is referred to as a followee or a friend of him. When a user posts some new content on Twitter, his followers can receive these messages on their own homepages and interact with him like forwarding (it's called retweet on Twitter) or replying to the tweets. Through subscribing to the content posted by others, a user can learn about what is happening with people and organizations he or she is interested in.

It has almost no barrier to post tweets on Twitter, people can share tweets about everything even trivial matters that traditional media won't pay attention to. Twitter is also famed for spreading messages almost instantly. For instance, an earthquake occurred in Morgan Hill, California on 30 March 2009. The first geocoded tweet about the earthquake arrived 19 s later, while the Northern California Seismic Network who has the Advanced National Seismic System cost 22 s to make an automatic response. Given all those factors, Twitter has become one of the main sources of acquiring information about real-life events. Twitter provides public Application Program Interfaces (APIs) for users to capture data which of interest. Users can collect real-time Twitter feeds or short-term historical tweets with specific keywords through these APIs. Using event detection techniques on Twitter data streams can help us to know which events are happening or happened recently, then we can consider for further analysis like the importance of events or the spatial and temporal patterns. Commonly, people show concern about real-time hot events or specific types of events, but here we pay more attention to events which are significant for particular individuals. For instance, suppose you are a fan of a celebrity, then you would like to know what happened to the celebrity or which events he is concerned about. In this article, we refer to these events are Personal Important Events (PIEs) for an individual. Individuals might not be involved in his PIEs (be one of the main characters of the events) but just participated in the discussion about those events. PIEs have a considerable effect on an individual, and it reflects on the change of the individual's local network structure. For example, a user shared a good viewpoint on an event and endorsed by many other users. Then he might get numerous new followers on Twitter and thus his local network structure changed sharply.

The local networks of users are always in an evolutionary process since there will be constantly new follow and unfollow behaviors. In this article, we study the dynamics of personal local networks of two Twitter accounts and observe how they evolve. We found that Twitter followership networks are highly dynamic. Within the one and a half months of our observation, about 9% and 22% of all connections have changed in the two personal local networks respectively. Most of the time, the evolution of personal local networks is in a steady state that new follows and unfollows come at a stable rate. However, when a PIE occurs, the violent disturbance is produced. In other words, PIEs cause bursts in the dynamics of the local network structure. We find that PIEs lead to two phenomena. One is that plenty of other users would link to the individual almost simultaneously, we refer to it as the follow burst. The other one is the individual's many followers drop the follow connections consecutively in a short time, and we call it the unfollow burst. Sometimes the two phenomena simultaneously occur on the individual. These bursts will significantly change the user's local network structure.

The remainder of this paper is organized as follows: We briefly review related works in Section 2. Then we describe our dataset and empirically study the evolution of users' personal social networks in Section 3. In Section 4 we introduce how to detect PIEs from user behaviors. We research on the effect of PIEs on users' local social network structure in Section 5 and give a simple sum-up in Section 6.

2. Related Works

Twitter event detection has always been a hot topic since Twitter launched in 2006. Researchers have shown that twitter event detection contributes to various application fields, such as epidemic diseases [1], political affairs [2–6], traffic conditions [7] and natural disaster emergencies [8–11]. Twitter event detection can be classified as specified and unspecified according to the event type.

Specified event detection uses pre-defined event information (e.g., keywords, hashtags) or known events. Lee and Sumuya [12] utilized the collective experiences and crowd behaviors on Twitter to detect geo-social events. Khurdiya et al. [13] proposed a framework based on Searching on Lucene with Replication (SOLR) and Conditional Random Field (CRF) which can identify small sub-events around a major event and build a map of them. Rill et al. [3] presented a system that uses special sentiment hashtags to detect emerging political events. Huang et al. [14] designed a high utility patter clustering framework that aims to detect and visualize small-scale city-level or even street-level events. Adedoyin–Olowe et al. [15] considered hashtags a significant and primary feature and used frequent pattern mining to capture word occurrence and detect sports and political events.

Unspecified event detection does not consider prior event information and mainly rely on bursty features, which is closer to our work. Abdelhaq et al. [16] developed a system called EvenTweet to detect localized events in real-time from a Twitter stream. Gao et al. [17] detected geographical social events by mining geographical temporal patterns and analyzing the content of tweets. Liu et al. [18] presented a system to detect burst events through mining burst words by incorporating features from message content, propagation periods, and other characteristics. Zhou and Chen [19] proposed a framework called Variable Dimensional Extendible Hash (VDEH) which fully utilizes the information of social data over multiple dimensions to detect composite social events over streams. Cheng and Wicks [20] used Space-Time Scan Statistics (STSS) technique to detect significant space-time events without using of tweet context. Alsaedi et al. [21] proposed an online combined classification-clustering framework to identify real-time events. Zhang et al. [22] introduced a graph-based event detection technique where nodes in the graph represent burst words and the edge weight is calculated according to their co-occurrence within each tweet. Strongly connected components in the graph are identified through the graph clustering technique that uses the depth-first search algorithm and the connected components are considered as events. Zhou et al. [23] developed a Bayesian model-based framework called Latent Event and Category Model (LECM) which assumes that each tweet is associated with an event. LECM extracts events from tweets and groups them into clusters with event type labels.

These event detection techniques utilize various kinds of features extracted from Twitter data, but new follow/unfollow count is rarely used. Since our work aims to detect the occurrence of events and does not focus on the details of events, we put forward a simple and efficient method which use new follow/unfollow count to detect events.

Except for proposing a new Twitter event detection method, we also analyze the dynamics of individuals' local followership networks. In fact, many works that target exclusively to the dynamics of online social networks have been done. In the early period, the emphasis of the works is on the modeling of various aspects of social network evolution over time [24–26]. More recently, the research hotspot has shifted to predict local changes in the network, such as the addition and deletion of specific edges between social actors [27–30]. In the process of predicting the edge creation and deletion, many features were found to be helpful, such as the network topological structure [31–33], the internal influence among the social actors [34–36], the external influences like other forms of media [37,38], and the nature of the information content itself [39–42]. Our work does not focus on the relevance between network dynamics and particular features, but stands in a macro perspective and identifies how events affect the evolution of social networks.

3. Brief Analysis of Twitter Followership Graph

In the section, we introduce our dataset which consists of multiple snapshots of two users' local social networks. We briefly analyzed the change of followers count, and further analysis of more features of Twitter followership networks will be done later.

3.1. Dataset Description

Our analysis focus on the users' local social networks. To keep things simple, we define a user's ego network as the subgraph made up of a user's followers (excluding the user himself) and all the follow relationships between them. The user himself is known as the ego and his followers are called alters.

Our dataset includes snapshots of two Twitter ego networks. The egos we chose are *@NZNationalParty* and *@nzlabour*, who are the official Twitter accounts of New Zealand Nation Party and New Zealand Labour Party separately. We observe the dynamics of the two networks from 10 September 2017 to 24 October 2017. This period of time just catches up the New Zealand general election. We fetch the data of the two egos' followers via the Twitter Representational State Transfer (REST) API. Twitter modified the access permission of REST API in 2013 and imposed a limit of 15 times requests to get the lists of followers each 15 min window for an authenticated user. This rate-limiting makes it harder to get a complete picture of the Twitter social graph. That is an important reason of the shifting for the research emphasis from the global Twitter network to the local Twitter network. The two ego networks have 14,306 and 23,877 nodes at the beginning respectively. We have recorded the exact timestamps of connections that created or deleted during the observation, hence we can study the fine-grained network dynamics.

3.2. Highly Dynamic of Twitter Network

Go through the everyday snapshots, we find that the Twitter ego networks are highly dynamic. Compare with the starting time, about 6.6% new connections came up in ego-network1 and 2.39% old links were deleted. Ego-network2 added about 19.1% new edges during the period, while 2.8% old links were removed. This shows the highly dynamic nature of Twitter followership networks. The removal of old edges accounts for a respectable proportion, which demonstrates that the thought of considering the Twitter graph as an 'only-growing' network in some previous works [27] is incorrect.

Figures 1 and 2 plot the counts of the everyday new follows and unfollows of @NZNationalParty and @nzlabour respectively. From Figures 1 and 2, we can discover that the churn rate of the followers of an individual remains consistent at most times. This steady background volatility gets interrupted when events happen, and then the network structure changes significantly.

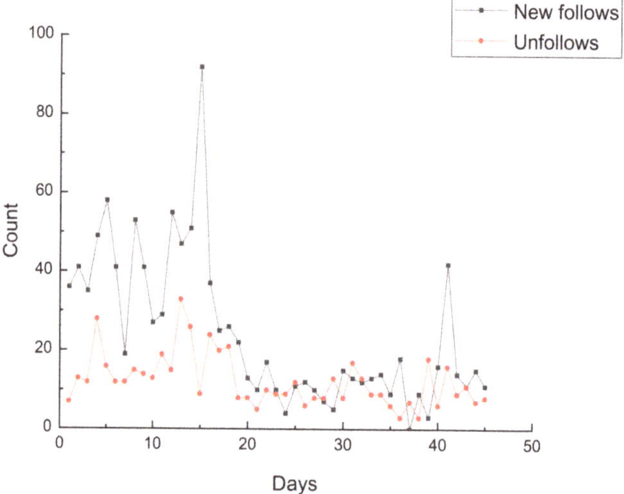

Figure 1. The variation of followers count of *@NZNationalParty*.

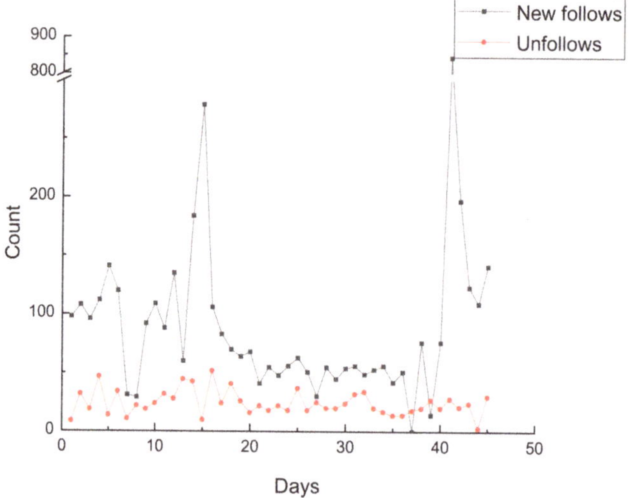

Figure 2. The variation of followers count of *@nzlabour*.

4. Personal Important Events Detection

4.1. Personal Important Events

A mass of events happens on Twitter every day, but most of them have few effects on a specific user's local network structure. Only if a user is referred to an event or participates in the discussion about the event, the event might provoke a great change of the users' local network structure. We define such events which significantly change a user's local network structure as Personal Important Events (PIEs) for the user.

We utilize the count variation of a user's followers to judge the PIEs for the users. That means, if a user experienced a follow burst or an unfollow burst in a time interval, we claim that a PIE for the user happened at that time. Then we try to discover PIEs of our observed objects, *@NZNationalParty* and *@nzlabour*. To detect the PIEs, or in other words, to identify the time intervals in which a user receives much more new follows or unfollows compared to what was expected historically, we process the data as follows.

4.2. The Bursts and the PIEs Detection

We use a method analogous to the way Myers and Leskovec used in [43] to detect bursts. We treat the arrival of new follows/unfollows of a user as an independent time series. We set the time interval as one day. Let $x = \{x_1, x_2, \ldots, x_n\}$ be the number of new follows/unfollows a user receives for each day. Let t_i denote the ith day of our observation period, and let $f(t_i)$ represent the difference between the actual new follows/unfollows and the expected value during t_i:

$$f(t_i) = x_i - E[x \mid t_i] = x_i - \frac{\sum_{j;\, 0 < t_i - t_j \leq 2} x_j \cdot \omega(t_i - t_j)}{\sum_{j;\, 0 < t_i - t_j \leq 2} \omega(t_i - t_j)} \tag{1}$$

There $\omega(t_i)$ is an exponentially decaying weight function. When the value of $f(t_i)$ is greater than a threshold, we consider a PIE happens at day t_i. In this article, we set the threshold as three standard deviations of the time series according to PauTa Criterion (3σ Criterion). We detected 5 PIEs for *@NZNationalParty* and 7 PIEs for *@nzlabour* during the observation period. We numbered the PIEs according to the time. As shown in Figure 3, the PIE 1~5 which are related to *@NZNationalParty* happened at Day4 (13 September 2017), Day12, Day14, Day15, and Day41 severally. Moreover, the PIE 6~12 related to *@nzlabour* happened at Day12, Day14, Day15, Day25, Day28, Day38, and Day41 separately. They are plotted on Figure 4. Refer to the news media, we find that PIEs are likely to correspond to real-world events about the New Zealand general election. We list the detected PIEs and the real-world events in Table 1 according to the dates.

Table 1. The correspondent relationship between Personal Important Events (PIEs) and real-world events.

Detected PIEs		Real-World Events
09.13.2017	PIE1	The National Party denounced the tax policy of the Labour Party.
09.21.2017	PIE2, PIE6	The final televised election debate was held.
09.23.2017	PIE3, PIE7	The general election was held.
09.24.2017	PIE4, PIE8	The preliminary result of electoral votes was announced.
10.04.2017	PIE9	The First Party prepared to negotiate with the National Party and the Labour Party.
10.07.2017	PIE10	The statistics for special votes was completed.
10.17.2017	PIE11	The leader of the Labour Party was suspected to hint that she's gonna win.
10.20.2017	PIE5, PIE12	The Labour Party won the election officially last night.

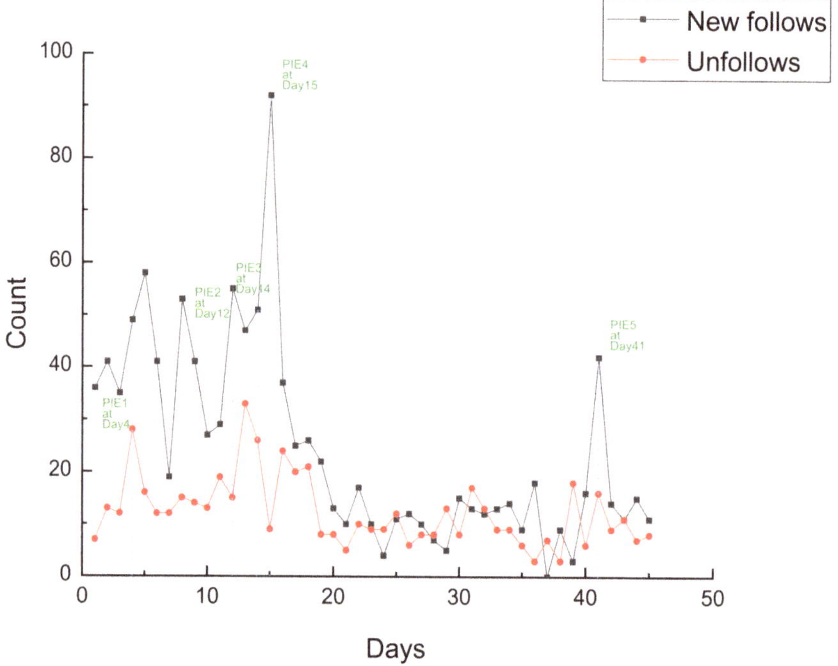

Figure 3. PIE 1–5 related to @*NZNationalParty*.

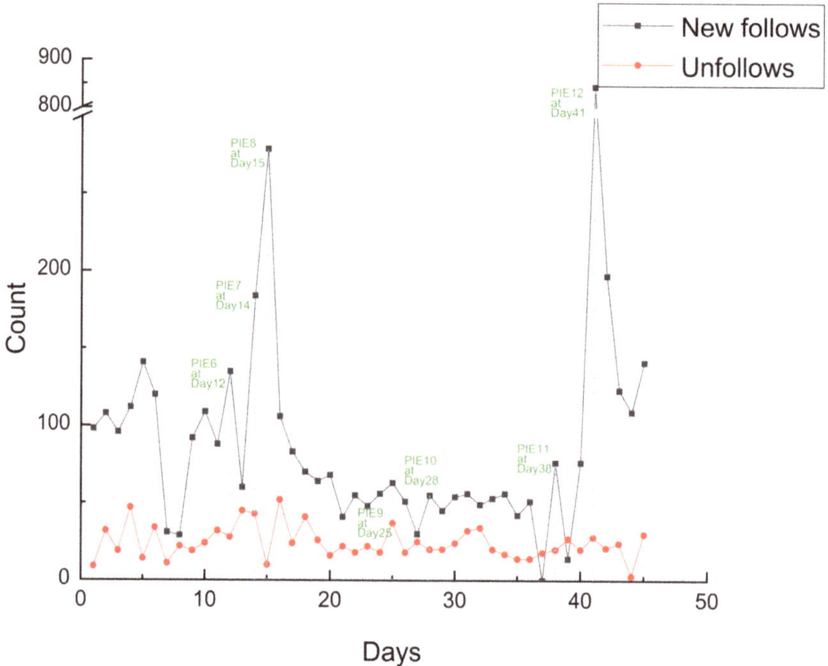

Figure 4. PIE 6–12 related to @*nzlabour*.

We classify PIEs into three categories according to the consequences brought to the corresponding user: follow-burst-PIEs for PIEs which lead to a follow burst, unfollow-burst-PIEs for PIEs which cause an unfollow burst, and mixed-burst-PIEs for PIEs result in both follow burst and unfollow burst simultaneously. Here we find that the PIE 1, 2, 4 for *@NZNationalParty* and the PIE 6, 8, 10 for *@nzlabour* are follow-burst-PIEs. The PIE 3 for *@NZNationalParty* and the PIE 7, 9 for *@nzlabour* are unfollow-burst-PIEs. The rest of PIEs including PIE 5, 11, 12 are mixed-burst-PIEs. The categories of PIEs are listed in Table 2. In a general way, a follow-burst-PIE for a given user is in favor of him in public opinion. On the contrary, an unfollow-burst-PIE for a user corresponds to an event which against him. The situation of a mixed-burst-PIE could be complicated since we are not sure whether a positive effect or a negative effect is brought to the related user.

Table 2. The categories of PIEs.

Related User of PIEs	*@NZNationalParty*	*@nzlabour*
Follow-burst-PIEs	PIE1, PIE2, PIE4	PIE6, PIE8, PIE10
Unfollow-burst-PIEs	PIE3	PIE7, PIE9
Mixed-burst-PIEs	PIE5	PIE11, PIE12

5. Evolution of Twitter Ego Networks

With the constant occurrence of link creation and deletion, the users' local network structure evolves as time goes by. In this section, we'll see how the two ego networks in our dataset evolve and how the bursts caused by PIEs effect on the evolution. Due to the complexity of mixed-burst-PIEs, here we only consider the effect of follow-burst-PIEs and unfollow-burst-PIEs on the ego network dynamics.

5.1. Follower Tweet Similarity

A user directly influences his followers (e.g., the members of his ego network) on Twitter through information diffusion. It's necessary to understand the relationship between a user and his followers. So we first explore how similar a user is to his followers, and how this similarity changes after the occurrence of a PIE for him. From a user's tweets (including retweets), we can basically know what he is interested in. For a pair of users, the textual similarity of their tweets would be a good indicator to quantify how similar their interests are. We aggregate the tweets posted by each user during the observation into single documents. Then we compute the cosine similarity of the Term Frequency-inverse Document Frequency (TF-IDF) weighted word vectors between two users' aggregated tweet documents and adopt it as user tweet similarity. First we extract all the key terms (assume that m terms in total) $W = \{w_1, w_2, ..., w_m\}$ from documents (of tweets). For a user x and his document (of tweets) d_x. The term frequency $tf_{(w_1, d_x)}$ measures how frequency term w_1 occurs in document d_x:

$$tf_{(w_1, d_x)} = \frac{\text{Number of times term } w_1 \text{ appears in document } d_x}{\text{Total number of terms in document } d_x} \tag{2}$$

The inverse document frequency $idf_{(w_1)}$ measures how important term w_1 is in all the documents:

$$idf_{(w_1)} = \ln \frac{\text{Total number of documents}}{\text{Number of documents with term } w_1 \text{ in it}} \tag{3}$$

Then the TF-IDF weight of term w_1 in document d_x is $tf\text{-}idf_{(w_1, d_x)} = tf_{(w_1, d_x)} \cdot idf_{(w_1)}$. Document d_x can be represented by a m dimension TF-IDF vector:

$$tf\text{-}idf_{(d_x)} = (tf\text{-}idf_{(w_1, d_x)}, tf\text{-}idf_{(w_2, d_x)}, ..., tf\text{-}idf_{(w_m, d_x)}) \tag{4}$$

In that way, the text similarity of user x and user y can be computed according to the cosine similarity between corresponding document TF-IDF vectors:

$$
\begin{aligned}
TextSim(x,y) &= \frac{tf\text{-}idf_{(d_x)} \cdot tf\text{-}idf_{(d_y)}}{\|tf\text{-}idf_{(d_x)}\| \cdot \|tf\text{-}idf_{(d_y)}\|} \\
&= \frac{\sum_{k=1}^{m} tf\text{-}idf_{(w_k,d_x)} \cdot tf\text{-}idf_{(w_k,d_y)}}{\sqrt{\sum_{k=1}^{m} tf\text{-}idf_{(w_k,d_x)}^{2}} \cdot \sqrt{\sum_{k=1}^{m} tf\text{-}idf_{(w_k,d_y)}^{2}}}
\end{aligned}
\tag{5}
$$

We define the follower tweet similarity of a user as the average value of the tweet similarity between a user and all his followers. By observing the follower tweet similarity of a user before and after a PIE, we investigate whether user's followers become more similar in their interests. We measured the follow tweet similarity of the egos for multiple days before and after the PIEs. To make the metric comparable across different users, we normalize each measurement by its value exactly at the day of the PIE and then average the metric across all PIEs of the same type. Figure 5 shows the result averaged across all follow-burst-PIEs and unfollow-burst-PIEs.

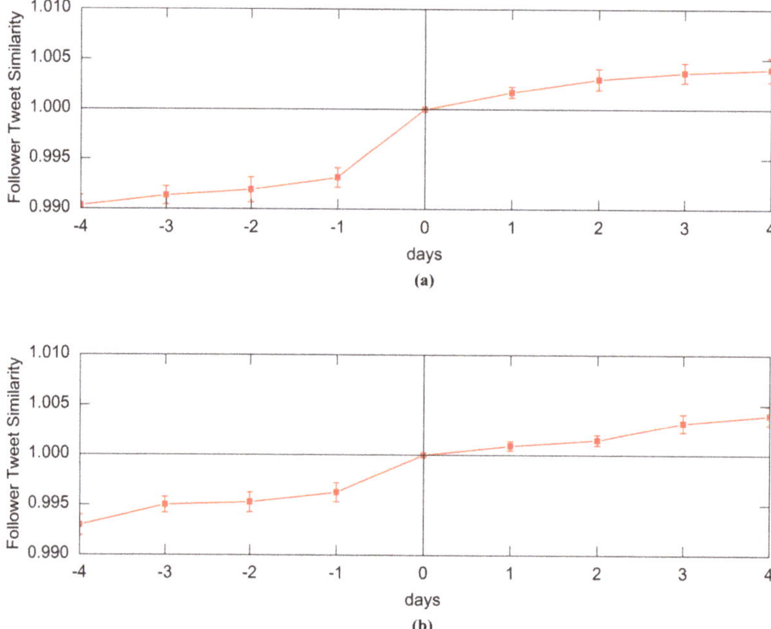

Figure 5. (**a**) Normalized Follower Tweet Similarity before and after a follow-burst-PIE; (**b**) Normalized Follower Tweet Similarity before and after an unfollow-burst-PIE.

We observe that the follower tweet similarity basically keeps increasing over time. This is easy to understand. When a user is new to Twitter, his earliest followers might not be that similar to him because they don't know his interests well enough. As the user's online behaviors increase, he is more likely to be followed by users who share common interests with him. We conclude that after whether a follow-burst-PIE or an unfollow-burst-PIE, the follower tweet similarity of ego networks obviously grows faster. A PIE makes the ego get more exposure on Twitter so users get to know more about his interests. Intuitively, a follow-burst-PIE attracts similar users to follow the ego, and an unfollow-burst-PIE prompt existing followers who are not similar to the ego to unfollow him. Both of them accelerate the rise of follower tweet similarity.

5.2. Follower Tweet Coherence

We discover that the followers become more in common with the ego after a PIE in the previous paragraph, then we are going to see if the followers become more related to one another (not just to the ego user). We use the same method of TF-IDF cosine similarity of tweet content to measure the similarity among the followers. Here we define the follower tweet coherence of a user as the average value of the tweet similarity of all pairs of his followers. We measure the tweet similarity across all pairs of followers of the ego in the days succeeding and preceding the PIEs. For large-scale networks who have too much node pairs, biased sampling strategies such as sample edge count [44] are recommended to calculate similarity between node pairs. The negative effect of sampling on similarity calculating is proved to be small [45]. These measurements are normalized by the value exactly on the day of the PIE. Moreover, we average the metric across all PIEs of the same type just as we did before. We plot the result in Figure 6.

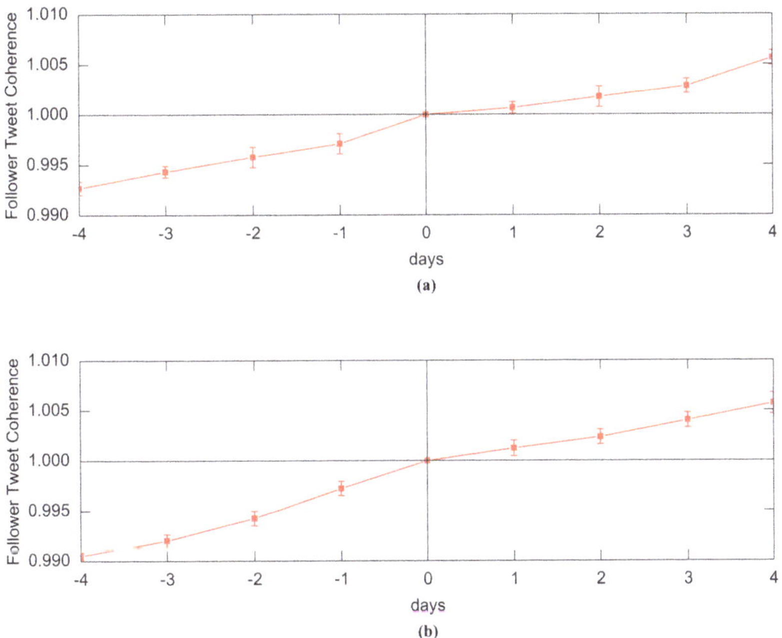

Figure 6. (**a**) Normalized Follower Tweet Coherence before and after a follow-burst-PIE; (**b**) Normalized Follower Tweet Coherence before and after an unfollow-burst-PIE.

A similar result to the follower tweet similarity is shown. Likewise, the follower tweet coherence increases over time, and speeds up after both follow-burst-PIEs and unfollow-burst-PIEs. Both types of PIEs cause the followers' interests and tweets to become more aligned with each other, make the user's ego network become more homogeneous. Since the followers become more similar to the ego as time goes by, it's predictable that followers become more alike among themselves. Combined with the result of follower tweet similarity, it's indicated that PIEs cause a process in the ego network's evolution toward bringing similar users together and pushing dissimilar users farther apart.

5.3. Connected Components Amongst Followers

After analyzing the similarity relationship between users before and after a PIE, the structural changes of a user's local neighborhood will be explored. In detail, we calculate the number of weakly connected components (WCC) of the ego networks during a burst. A weakly connected component of a directed graph is a maximum subgraph in which any two nodes are connected by direct edge path. For any user in a weakly connected component of a Twitter followership network, there exists at least one another user follows or followed by him. If the number of WCC of an ego network is high, that means the subgraph of the user's followers is fragmented. It indicates that user's followers tend not to follow each other and do not belong to a single cohesive community. After measuring the number of WCC in the days before and after the PIEs, we execute the normalization and equalization like before.

Figure 7 shows the relative number of WCC in the days preceding and succeeding PIEs. We discover that there is an upward trend in the number of WCC. That means the followers are divided into more fragmented communities, though the followers become more similar over time. It's not hard to explain. The followers of a user increase over time since new followers arrive faster than old followers leave in ordinary times. While these new followers have parallel interests, there is no enough time for them to know each other and follow each other (they are likely to follow each other after a period of time but not immediately). Hence, there would be more weakly connected components of Twitter ego networks.

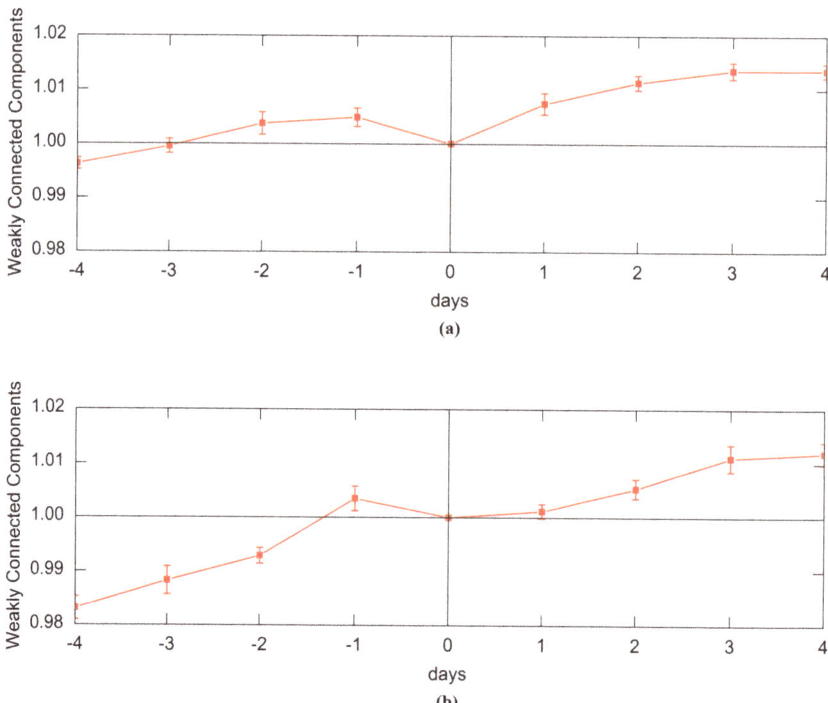

Figure 7. (a) The number of normalized Weakly Connected Components before and after a follow-burst-PIE; (b) The number of normalized Weakly Connected Components before and after an unfollow-burst-PIE.

This upward trend of the number of WCC is interrupted by PIEs. Whether a follow-burst-PIE or an unfollow-burst-PIE decreases the number of WCC. A PIE prompts users to post tweets to discuss the event, and this speeds up the process of users knowing each other. While a follow-burst-PIE brings numerous new members to the ego network, the newcomers may be familiar with old ones in a short time and follow each other. The number of WCC, therefore, will not increase after a follow burst-PIE. On the other hand, an unfollow-burst-PIE expels those users who are not close to the others out of the ego network. The ego network then becomes more tightly connected and the number of WCC gets reduced.

5.4 Followers Following Each Other

Lastly, we focus on the edge density of the ego networks. For a given user, the metric represents what fraction of all possible following relationships between his followers actually exist. Edge density measures the degree that a user's followers tend to follow each other. A lower value of the number of WCC indicates that information can spread to a broader range, while a higher value of edge density means a faster propagation speed in a local scope. Similarly, we measure the edge density of ego networks in the days before and after the PIEs. Normalized treatment and averaging treatment are given then.

Since users' ego networks always keep growing in the number of nodes, we guess that the edge density of ego networks declines over time. Figure 8 confirms our thought. We notice that there is a steady decrease in edge density before either type of PIEs. For the days after a burst, however, something interesting happens. For the unfollow-burst-PIEs, the density increases, while for the follow- burst-PIEs, the density still decreases but more slowly. We explain the two observations as follows. A follow-burst-PIE will bring plentiful new followers to the ego. While the newcomers and old followers will follow each other within a short time, as we mentioned above, the newly-established follow connections won't be too many. Thus the edge density will still decrease but slower. On the other hand, an unfollow-burst-PIE won't bring too many new followers but expel users who are loosely connected with others. Therefore, the edge density rises after an unfollow-burst-PIE. In general, both follow-burst-PIEs and unfollow-burst-PIEs inhibit the downtrend of the edge density.

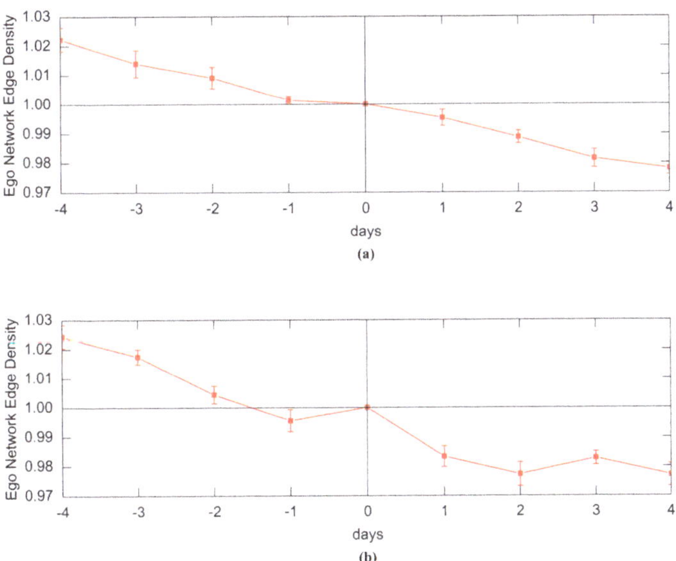

Figure 8. (**a**) Normalized Edge Density before and after a follow-burst-PIE; (**b**) Normalized Edge Density before and after an unfollow-burst-PIE.

6. Conclusions

In this paper, we propose an event detection method to detect the occurrences of significant events for specific individuals. This method only based on the variation of the user's followers is quite simple but effective. We divide the events into categories according to the positive or negative effect on the particular user. Further, we observe the evolution of individuals' Twitter followership networks and see if different types of events have different influences on the network dynamics. On some of the features, events accelerate the original evolutionary trend. In other features, events suppress the original trend.

Understanding the evolution trend of Twitter followership networks and its reactions to events are helpful to investigate further on Twitter followership networks. We can control the occurrence of events to make target networks achieve a desired state. This can be applied in various fields such as public opinion monitoring, disaster warning, crisis management, and intelligent decision making. Besides, our work in this paper is limited to the analysis of Twitter followership networks. Further works can be done for the research of network dynamics on more kinds of Twitter networks such as retweet networks and mention networks.

Author Contributions: Conceptualization, T.T. and G.H.; methodology, T.T.; software, T.T.; validation, T.T.; formal analysis, T.T.; investigation, T.T.; resources, T.T.; data curation, T.T.; writing—original draft preparation, T.T.; writing—review and editing, T.T.; visualization, T.T.; supervision, G.H.; project administration, G.H.; funding acquisition, T.T. All authors have read and agreed to the published version of the manuscript.

Funding: This research was funded by the National Natural Science Foundation of China grant number 0561701074.

Conflicts of Interest: The authors declare no conflict of interest.

Abbreviations

The following abbreviations are used in this manuscript:

API	Application Program Interface
PIE	Personal Important Event
SOLR	Searching on Lecene with Replication
CRF	Conditional Random Field
VDEH	Variable Dimensional Extendible Hash
STSS	Space-Time Scan Statistics
LECM	Latent Event and Category Model
REST	Representational State Transfer
TF-IDF	Term Frequency-inverse Document Frequency
WCC	Weakly Connected Components

References

1. Lampos, V.; Cristianini, N. Nowcasting events from the social web with statistical learning. *ACM Trans. Intell. Syst. Technol. (TIST)* **2012**, *3*, 1–22. [CrossRef]
2. Mitchell, A.; Hitlin, P. Twitter Reaction to Events Often at Odds with Overall Public Opinion. 2013. Avaliable online: https://apo.org.au/node/33069 (accessed on 16 September 2020).
3. Rill, S.; Reinel, D.; Scheidt, J.; Zicari, R.V. Politwi: Early detection of emerging political topics on twitter and the impact on concept-level sentiment analysis. *Knowl. Based Syst.* **2014**, *69*, 24–33. [CrossRef]
4. Nguyen, D.T.; Jung, J.E. Real-time event detection for online behavioral analysis of big social data. *Future Gener. Comput. Syst.* **2017**, *66*, 137–145. [CrossRef]
5. Saeed, Z.; Abbasi, R.A.; Sadaf, A.; Razzak, M.I.; Xu, G. Text stream to temporal network-a dynamic heartbeat graph to detect emerging events on Twitter. In *Pacific-Asia Conference on Knowledge Discovery and Data Mining*; Springer: Berlin/Heidelberg, Germany, 2018; pp. 534–545.
6. Saeed, Z.; Abbasi, R.A.; Razzak, M.I.; Xu, G. Event detection in Twitter stream using weighted dynamic heartbeat graph approach. *arXiv* **2019**, arXiv:1902.08522.
7. Ribeiro, S.S., Jr.; Davis, C.A., Jr.; Oliveira, D.R.R.; Meira, W., Jr.; Gonçalves, T.S.; Pappa, G.L. Traffic observatory: A system to detect and locate traffic events and conditions using Twitter. In Proceedings of the 5th ACM SIGSPATIAL International Workshop on Location-Based Social Networks, Redondo Beach, CA, USA, 6 November 2012; pp. 5–11.
8. Sakaki, T.; Okazaki, M.; Matsuo, Y. Earthquake shakes Twitter users: Real-time event detection by social sensors. In Proceedings of the 19th International Conference on World Wide Web, Raleigh, NC, USA, 26–30 April 2010; pp. 851–860.

9. Vieweg, S.; Hughes, A.L.; Starbird, K.; Palen, L. Microblogging during two natural hazards events: What twitter may contribute to situational awareness. In Proceedings of the SIGCHI Conference on Human Factors in Computing Systems, Atlanta, GA, USA, 10–15 April 2010; pp. 1079–1088.

10. Takahashi, T.; Abe, S.; Igata, N. Can twitter be an alternative of real-world sensors? In *International Conference on Human-Computer Interaction*; Springer: Berlin/Heidelberg, Germany, 2011; pp. 240–249.

11. Crooks, A.; Croitoru, A.; Stefanidis, A.; Radzikowski, J. # Earthquake: Twitter as a distributed sensor system. *Trans. GIS* **2013**, *17*, 124–147.

12. Lee, R.; Sumiya, K. Measuring geographical regularities of crowd behaviors for Twitter-based geo-social event detection. In Proceedings of the 2nd ACM SIGSPATIAL International Workshop on Location Based Social Networks, San Jose, CA, USA, 3–5 November 2010; pp. 1–10.

13. Khurdiya, A.; Dey, L.; Mahajan, D.; Verma, I. Extraction and Compilation of Events and Sub-events from Twitter. In Proceedings of the 2012 IEEE/WIC/ACM International Conferences on Web Intelligence and Intelligent Agent Technology, Macau, China, 4–7 December 2012; Volume 1, pp. 504–508.

14. Huang, J.; Peng, M.; Wang, H. Topic detection from large scale of microblog stream with high utility pattern clustering. In Proceedings of the 8th Workshop on Ph.D. Workshop in Information and Knowledge Management, Melbourne, Australia, 19 October 2015; pp. 3–10.

15. Adedoyin-Olowe, M.; Gaber, M.M.; Dancausa, C.M.; Stahl, F.; Gomes, J.B. A rule dynamics approach to event detection in twitter with its application to sports and politics. *Expert Syst. Appl.* **2016**, *55*, 351–360. [CrossRef]

16. Abdelhaq, H.; Sengstock, C.; Gertz, M. Eventweet: Online localized event detection from twitter. *Proc. VLDB Endow.* **2013**, *6*, 1326–1329. [CrossRef]

17. Gao, X.; Cao, J.; He, Q.; Li, J. A novel method for geographical social event detection in social media. In Proceedings of the Fifth International Conference on Internet Multimedia Computing and Service, Huangshan, China, 17–18 August 2013; pp. 305–308.

18. Liu, C.; Xu, R.; Gui, L. Burst events detection on micro-blogging. In Proceedings of the 2013 International Conference on Machine Learning and Cybernetics, Tianjin, China, 14–17 July 2013; Volume 4, pp. 1921–1924.

19. Zhou, X.; Chen, L. Event detection over twitter social media streams. *VLDB J.* **2014**, *23*, 381–400. [CrossRef]

20. Cheng, T.; Wicks, T. Event detection using Twitter: A spatio-temporal approach. *PLoS ONE* **2014**, *9*, e97807. [CrossRef] [PubMed]

21. Alsaedi, N.; Burnap, P.; Rana, O.F. A combined classification-clustering framework for identifying disruptive events. In Proceedings of the ASE SocialCom Conference, Stanford, CA, USA, 27–31 May 2014.

22. Zhang, X.; Chen, X.; Chen, Y.; Wang, S.; Li, Z.; Xia, J. Event detection and popularity prediction in microblogging. *Neurocomputing* **2015**, *149*, 1469–1480.

23. Zhou, D.; Chen, L.; He, Y. An unsupervised framework of exploring events on twitter: Filtering, extraction and categorization. In Proceedings of the Twenty-Ninth AAAI Conference on Artificial Intelligence, Austin, TX, USA, 25–30 January 2015.

24. Backstrom, L.; Huttenlocher, D.; Kleinberg, J.; Lan, X. Group formation in large social networks: Membership, growth, and evolution. In Proceedings of the 12th ACM SIGKDD International Conference on Knowledge Discovery and Data Mining, Philadelphia, PA, USA, 20–23 August 2006; pp. 44–54.

25. Leskovec, J.; Backstrom, L.; Kumar, R.; Tomkins, A. Microscopic evolution of social networks. In Proceedings of the 14th ACM SIGKDD International Conference on Knowledge Discovery and Data Mining, Las Vegas, NV, USA, 24–27 August 2008; pp. 462–470.

26. Kumar, R.; Novak, J.; Tomkins, A. Structure and evolution of online social networks. In *Link Mining: Models, Algorithms, and Applications*; Springer: Berlin/Heidelberg, Germany, 2010; pp. 337–357.

27. Backstrom, L.; Leskovec, J. Supervised random walks: Predicting and recommending links in social networks. In Proceedings of the Fourth ACM International Conference on Web Search and Data Mining, Hong Kong, China, 9–12 February 2011.

28. Kwak, H.; Moon, S.; Lee, W. More of a receiver than a giver: Why do people unfollow in Twitter? In Proceedings of the Sixth International AAAI Conference on Weblogs and Social Media, Dublin, Ireland, 4–7 June 2012.

29. Xu, B.; Huang, Y.; Kwak, H.; Contractor, N. Structures of broken ties: Exploring unfollow behavior on twitter. In Proceedings of the 2013 Conference on Computer Supported Cooperative Work, San Antonio, TX, USA, 23–27 February 2013; pp. 871–876.

30. Hutto, C.J.; Yardi, S.; Gilbert, E. A longitudinal study of follow predictors on twitter. In Proceedings of the Sigchi Conference on Human Factors in Computing Systems, Paris, France, 27 April–2 May 2013; pp. 821–830.

31. Kempe, D.; Kleinberg, J.; Tardos, É. Maximizing the spread of influence through a social network. In Proceedings of the Ninth ACM SIGKDD International Conference on Knowledge Discovery and Data Mining, Washington, DC, USA, 24–27 August 2003; pp. 137–146.

32. Ugander, J.; Backstrom, L.; Marlow, C.; Kleinberg, J. Structural diversity in social contagion. *Proc. Natl. Acad. Sci. USA* **2012**, *109*, 5962–5966. [CrossRef]

33. Romero, D.M.; Tan, C.; Ugander, J. On the interplay between social and topical structure. In Proceedings of the Seventh International AAAI Conference on Weblogs and Social Media, Cambridge, MA, USA, 8–11 July 2013.

34. Bakshy, E.; Karrer, B.; Adamic, L.A. Social influence and the diffusion of user-created content. In Proceedings of the 10th ACM Conference on Electronic Commerce, Stanford, CA, USA, 6–10 July 2009; pp. 325–334.

35. Cha, M.; Haddadi, H.; Benevenuto, F.; Gummadi, K.P. Measuring user influence in twitter: The million follower fallacy. In Proceedings of the Fourth International AAAI Conference on Weblogs and Social Media, Washington, DC, USA, 23–26 May 2010.

36. Bakshy, E.; Hofman, J.M.; Mason, W.A.; Watts, D.J. Everyone's an influencer: Quantifying influence on twitter. In Proceedings of the Fourth ACM International Conference on Web Search and Data Mining, Hong Kong, China, 9–12 February 2011; pp. 65–74.

37. Crane, R.; Sornette, D. Robust dynamic classes revealed by measuring the response function of a social system. *Proc. Natl. Acad. Sci. USA* **2008**, *105*, 15649–15653. [CrossRef] [PubMed]

38. Myers, S.A.; Zhu, C.; Leskovec, J. Information diffusion and external influence in networks. In Proceedings of the 18th ACM SIGKDD International Conference on Knowledge Discovery and Data Mining, Beijing, China, 12–16 August 2012; pp. 33–41.

39. Berger, J.; Milkman, K. Social transmission, emotion, and the virality of online content. *Whart. Res. Pap.* **2010**, *106*, 1–52.

40. Hong, L.; Dan, O.; Davison, B.D. Predicting popular messages in twitter. In Proceedings of the 20th International Conference Companion on World Wide Web, Hyderabad, India, 28 March–1 April 2011; pp. 57–58.

41. Naveed, N.; Gottron, T.; Kunegis, J.; Alhadi, A.C. Bad news travel fast: A content-based analysis of interestingness on twitter. In Proceedings of the 3rd International Web Science Conference, Koblenz, Germany, 15–17 June 2011; pp. 1–7.

42. Tsur, O.; Rappoport, A. What's in a hashtag? Content based prediction of the spread of ideas in microblogging communities. In Proceedings of the Fifth ACM International Conference on Web Search and Data Mining, Seattle, WA, USA, 8–12 February 2012; pp. 643–652.

43. Myers, S.A.; Leskovec, J. The bursty dynamics of the twitter information network. In Proceedings of the 23rd International Conference on World Wide Web, Seoul, Korea, 7–11 April 2014; pp. 913–924.

44. Maiya, A.S.; Berger-Wolf, T.Y. Benefits of bias: Towards better characterization of network sampling. In Proceedings of the 17th ACM SIGKDD International Conference on Knowledge Discovery and Data Mining, San Diego, CA, USA, 21–24 August 2011; pp. 105–113.

45. Ahmed, N.K.; Duffield, N.; Neville, J.; Kompella, R. Graph sample and hold: A framework for big-graph analytics. In Proceedings of the 20th ACM SIGKDD International Conference on Knowledge Discovery and Data Mining, New York, NY, USA, 24–27 August 2014; pp. 1446–1455.

 *information*

Article

Social Capital on Social Media—Concepts, Measurement Techniques and Trends in Operationalization

Flora Poecze [1] **and Christine Strauss** [2,*]

[1]   Institute of Information Systems, Vienna University of Technology, 1040 Vienna, Austria;
      flora.poecze@tuwien.ac.at
[2]   Department of Marketing and International Business, University of Vienna, 1010 Vienna, Austria
*   Correspondence: christine.strauss@univie.ac.at

Received: 27 September 2020; Accepted: 2 November 2020; Published: 4 November 2020

Abstract: The introduction of the Web 2.0 era and the associated emergence of social media platforms opened an interdisciplinary research domain, wherein a growing number of studies are focusing on the interrelationship of social media usage and perceived individual social capital. The primary aim of the present study is to introduce the existing measurement techniques of social capital in this domain, explore trends, and offer promising directions and implications for future research. Applying the method of a scoping review, a set of 80 systematically identified scientific publications were analyzed, categorized, grouped and discussed. Focus was placed on the employed viewpoints and measurement techniques necessary to tap into the possible consistencies and/or heterogeneity in this domain in terms of operationalization. The results reveal that multiple views and measurement techniques are present in this research area, which might raise a challenge in future synthesis approaches, especially in the case of future meta-analytical contributions.

Keywords: social capital; social media; operationalization; measurement; scoping review

---

1. Introduction

The launch of Web 2.0 at the turn of the 21st century enabled a communication revolution. This was followed by the rapid emergence of diverse social media platforms, of which Friendster was one of the first globally known ones; in turn, a growing scientific interest started to characterize the present era [1,2]. This concentrated attention brought up a heterogeneous set of terminological approaches for the novel phenomenon termed "social media" (SM) [3].

Scientific publications in this domain commonly highlight the interactive function of the platforms in question. Furthermore, their services that offer instant communication, extended with possibilities offered as user-generated content (UGC) such as liking, sharing, and commenting. Based on such reasons, one of the most widely used definitions in this area of research is offered by Kaplan and Haenlein (2010), according to whom social media platforms are "internet-based applications that build on the ideological and technological foundations of Web 2.0, and that allow the creation and exchange of user generated content" [4]. VanMeter et al. (2015) stated that such platform is an interactive one, which "allows social actors to create and share in multi-way, immediate and contingent communications" [3]. The core nature of social media platforms is the purpose of enhancement and maintenance of individual user relationships; therefore, SM use can be considered as an investment in social relationships [5]. The exponential user base growth on a plethora of social media platforms has a multitude of individual, underlying reasons, from which this paper will tap into the human drive for social interactions and engagement, and evolutionary phenomena (e.g., survival, reproduction), facilitated by cooperation and trust, leading to the perception of an individual's reputation.

The human need to belong is a widely discussed psychological phenomenon [6]. There are empirical suggestions for the influence of this drive on college students' social media use, manifesting in interaction and social engagement [7]. The need for belongingness appeared in Maslow's [8] theory of human motivation, which is manifesting itself on today's social media platforms. Drawing back to the thoughts of Baumeister and Leary (1995), this motivation leads individuals to enhanced efforts to broaden and strengthen their social connections. As discussed in the previous social media literature, belongingness and self-representation [9] are considered among the primary reasons for social media use [10]. Taking into of Lin's (2001) definition, social capital is the "investment in social relations with expected returns in the marketplace" [11]. Under these circumstances, individual social media presence is indeed influenced by the drive for belongingness, paired with expected returns, which, under these circumstances, indicates enhanced social ties and the possibility of leveraging social support [12].

Considering the exponential growth of the global user base and the diversity of social media platforms, which have become part of humanity's everyday lives, the measurement of individual social capital on social media has become a crucial area of investigation. The expected and perceived returns, which characterize this phenomenon make it essential for scientists to empirically examine and quantify their potential impacts on the lives of individuals at a global scale. Therefore, emerging scientific attention has turned to operationalize and validate social capital scales, the evaluation of which can possibly describe this phenomenon in detail.

There are indications, however, that there are possible inconsistencies present regarding the measurement techniques of perceived individual social capital. Williams (2006) pointed out, based on the theories of Putnam (2000), that the bridging and bonding dimensions of the cultural view of social capital are not orthogonal (cf. ibid., pp. 596–597), which can lead to possible measurement discrepancies. Further concerns suggested that the distinction between bridging and bonding social capital is rather ad hoc [13]; therefore, their treatment in such manner can possibly lead to harmful consequences [14]. Additional critics [14] highlighted the issue that Granovetter's seminal works about the implications of weak and strong social ties [15,16], which indicated that weak ones are possibly more important than strong ones, are dominating this research field [17].

In this manner, this investigation aims to scope the presently existing measurement techniques and trends in this research area. No hypotheses were set in this study, reflecting on its exploratory nature. The rest of this article is structured as follows. In the next section, we present a terminological outline of the existing social capital theories. This leads to differentiation regarding online and offline social capital. The birth and theoretical development of this phenomenon is then presented. This includes a discussion of the cultural and multidimensional views of social capital, as subjects of the vast majority of publications operationalizing social capital. The items of measurement are detailed. This is followed by a presentation of the research methodology, aiming to explore past scientific research measuring social capital, paying attention to possible inconsistencies regarding operationalization techniques. The results are presented next, followed by the conclusions, limitations of the present paper, and future research suggestions.

2. Literature Review

Social capital terminology has undergone a substantial transformation since its birth. As a possible reason for such diversity, Fine (2010) [18] highlighted that previous research practices applied an appropriate definition matching the particular application in question [19–22]. According to Bourdieu, social capital is the "aggregate of the actual or potential resources which are linked to possession of a durable network of more or less institutionalized relationships of mutual acquaintance and recognition" [23], emphasizing that it is a form of capital, measurable on an individual or group level, characterizing embedded relationships between individuals. Furthermore, Coleman defines social capital as an accumulation of resources stemming from various individual relationships [24].

Portes highlighted the importance of the structure of such relationships, wherein the actors of this phenomenon are located [22], while Fukuyama underlined the importance of co-operation among individuals, which promotes social capital [25].

According to Putnam, social capital is social networks associated with the norms of reciprocity, indicating that the phenomenon itself jointly describes these networks and their effects on participating individuals. [26] However, to offer a brief outlook regarding the up to date inconsistencies regarding social capital discussed in detail by Fine [18], the argument of social capital being the cause, the effect, or the process itself, is a matter of present scientific debates as well [27].

The emergence of Web 2.0 brought further developments regarding the theory and parallel development of the applied measurement techniques of social capital, distinguishing between online and offline contexts, with research indicating that the use of the Internet is associated with trust and community involvement enhancement. [28] The impact of the Internet as a surrogate and a supplement of human communication has been discussed widely in scientific research, the focus of which has been e-mail usage [29], or the functions of chat rooms in idea sharing and political participation [30] in the early stages of Internet studies in social capital research.

With the emergence of computer-mediated social networks, the discussion of the associations between social capital and individual tie strength research, investigating both strong and weak ties in an online context based on Granovetter's social tie theories [15,16], opened a new research field. The aim for the development of online social capital became one of the core aspects of empirical research, a milestone of which was the development and validation of the first comprehensive online social capital scale by Williams [27]. To better understand the importance and details of this contribution, it is essential to discuss the most widely discussed, existing views of social capital theory. This is crucial to grasp the context of this scale, which still has one of the most noted impacts in the present, empirical social capital research measured on social media.

The birth of social capital as a scientific phenomenon is unclear [18]. According to Hofer and Aubert (2013), Lyda Hanifan's article from 1916 [31] can be seen as a possible theoretical root [32]. Hanifan's rediscovery during the beginning of the 21st century can be potentially attributed to an article by Putnam and Goss (2002), in which the authors stated that her definition encompassed all of the crucial elements identified ones in contemporary science [33].

The beginning of the 1980s marked its first concentrated attention through the works of French radical sociologist Pierre Bourdieu, joined by American rational choice sociologist James Coleman, who started elaborating on this topic in the late 1980s and early 1990s [34]. Robert Putnam's investigations during the turn of the century [26] took an important scientific step towards, through the definition and conceptualization of bridging and bonding social capital. This based on Granovetter's works on social tie strength [15,16], wherein he proposed that strong social ties (e.g., family or friends) are not valuable for an individual in the process of a new job acquisition. However, weak social ties (i.e., the vast network of acquaintances) are beneficial for the individual in question [15]. Based on Granovetter's seminal works, Putnam (2000) proposed that a person's bridging social capital (i.e., weak social ties) is valuable for the acquisition of previously unknown, new information, while the function of bonding social capital refer to an individual's strongest ties, is the provision of social and emotional support (cf., [35]).

In parallel to the previously described cultural view of social capital [23,24,26], and theorists of the structural view [3,8,18], the turn of the century marked Nahapiet and Goshal's seminal work (1998). This work elaborated on the multidimensional view of social capital, segmenting it into structural (i.e., social interaction ties), relational (i.e., shared language, cultural understanding) and cognitive (i.e., trust, norms, obligations, identification) dimensions [36].

Social capital has been put into relationship analyses with enormously diverse phenomena. Fine (2010) offered several curious examples [18] (e.g., the prevention of deforestation [37], skin color as a factor in marriage prospects [38], or pets as social capital conduits [39]). The most promising research platforms in social capital studies are social media platforms, building upon belongingness as a human drive, and social engagement. The shared question of such studies is, whether or not the use of social media affects the individual's perception of the social capital of the self and the perceived social support (see the meta-analysis of Domahidi, 2018 [40]). As a result, the number of papers is growing rapidly, in which the perceived individual social capital is analyzed on social media platforms [41].

The different aspects, viewpoints, and theoretical considerations in terms of social capital raise the question of how these are operationalized for further empirical investigation and evidence. The development of measurement techniques was already urged by Quan-Haase and Wellmann in 2004, who argued for its necessity based on the accelerated emergence of the Internet in parallel to the development of social capital [42]. Following this call, Williams (2006) created the Internet Social Capital Scale (ISCS), consisting of two scales proposed to measure bridging, and bonding social capital, respectively, including 10-10 measurement items, based on Putnam's (2000) conceptualization. [27] These scales were extended and modified [1], wherein the bridging and bonding social capital of Michigan State University (MSU) students was measured. The aim was the analysis of student social capital, the intensity of Facebook use and further control variables. In this article, the authors introduced the definition and measurement of maintained social capital as well, which refers to such prior, high school social connections of students that were later maintained during their time spent in higher education.

The following table aims to introduce the measurement items of the ISCS [27], in comparison to those items that were included in the seminal work of Ellison et al. (2007) (cf. Table 1).

Table 1 illustrates that Ellison et al. (2007) adapted five statements from the ISCS, slightly adjusting the statements to the MSU context, while they operationalized and validated a scale for maintained social capital as well. Therefore, Table 1 presents the five distinct measurement items of this social capital construct as well.

The dichotomous handling of bridging and bonding social capital has raised concerns. As Williams (2006) pointed out, these constructs "are not mutually exclusive, [. . .], they are oblique rather than orthogonal to one another" [27]. Their treatment as distinctive constructs can result in harmful consequences; therefore, they should be handled as oblique ones [14]. Further critics noted that the distinction between bridging and bonding social capital measurement instruments is rather ad hoc [13]. Additionally, based on the wide recognition of Granovetter (1973), which highlighted the importance of weak social ties, academic research tended to highlight the existence of this phenomenon and sought evidence for its underpinning [17]. This has also generated concerns in recent studies [14].

The development of the measurement constructs in the multidimensional view of social capital followed a different path in operationalization (see [4], pp. 140–157, for a critical summary) from the end of the 1990s until 2006. This marked the year of the publication of the article by Chiu et al. (2006), which created and validated a comprehensive set of items for all three studied social capital dimensions (i.e., structural, relational, cognitive) (Table 2) [43].

The history of measurement development in the multidimensional view of social capital reached a milestone with the seminal work of Chiu et al. (2006), which built upon the work of Tsai and Goshal (1998). The definition of the structural dimension as social interaction ties, the relational one as trustworthiness and trust, and the cognitive one as shared vision (in an enterprise setting) was defined by these authors. They created a standardized betweenness index for the evaluation of social interaction ties, while standardized in-degree, centrality was calculated for the measurement of trust and trustworthiness. The cognitive dimension was measured through two Likert-scale items [44].

Table 1. Comparison of measurement items from Williams' Internet Social Capital Scale (ISCS) with measurement items from Ellison et al. (2007). MSU: Michigan State University.

Online/Offline Bridging SC [27]	Online/Offline Bonding SC [27]	Bridging SC [1]	Bonding SC [1]	Maintained SC [1]
Interacting with people online/offline makes me interested in things that happen outside my town	There are several people online/offline I trust to help solve my problems	I feel I am part of the MSU community	There are several people at MSU I trust to solve my problems	I'd be able to find out about events in another town from a high school acquaintance living there
Interacting with people online/offline makes me want to try new things	There is someone online/offline I can turn to for advice about making very important decisions	I am interested in what goes on at MSU	If I needed an emergency loan of $100, I know someone at MSU I can turn to	If I needed to, I could ask a high school acquaintance to do a small favor for me
Interacting with people online/offline makes me interested in what people unlike me are thinking	There is no one online/offline that I feel comfortable talking to about intimate personal problems. (reversed)	MSU is a good place to be	There is someone at MSU I can turn to for advice about making very important decisions	I'd be able to stay with a high school acquaintance if traveling to a different city
Talking with people online/offline makes me curious about other places in the world	When I feel lonely, there are several people online/offline I can talk to.	I would be willing to contribute money to MSU after graduation	The people I interact with at MSU would be good job references for me	I would be able to find information about a job or internship from a high school acquaintance
Interacting with people online/offline makes me feel like part of a larger community	If I needed an emergency loan of $500, I know someone online/offline I can turn to.	Interacting with people at MSU makes me want to try new things	I do not know people at MSU well enough to get them to do anything important (reversed)	It would be easy to find people to invite to my high school reunion
Interacting with people online/offline makes me feel connected to the bigger picture	The people I interact with online/offline would put their reputation on the line for me.	Interacting with people at MSU makes me feel like a part of a larger community		
Interacting with people online/offline reminds me that everyone in the world is connected	The people I interact with online/offline would be good job references for me.	I am willing to spend time to support general MSU activities		
I am willing to spend time to support general online/offline community activities	The people I interact with online/offline would share their last dollar with me.	At MSU, I come into contact with new people all the time		
Interacting with people online/offline gives me new people to talk to	I do not know people online/offline well enough to get them to do anything important. (reversed)	Interacting with people at MSU reminds me that everyone in the world is connected		
Online/Offline, I come in contact with new people all the time	The people I interact with online/offline would help me fight an injustice.			

Table 2. Social capital dimensions, subscales and respective measurement items based on Chiu et al. (2006).

Social Capital Dimension	Subscale	Measurement Item Number	Measurement Item Statement
structural dimension	social interaction ties	1.	I maintain close social relationships with some members in the BlueShop virtual community.
		2.	I spend a lot of time interacting with some members in the BlueShop virtual community.
		3.	I know some members in the BlueShop virtual community on a personal level.
		4.	I have frequent communication with some members in the BlueShop virtual community.
relational dimension	trust	1.	Members in the BlueShop virtual community will not take advantage of others even when the opportunity arises.
		2.	Members in the BlueShop virtual community will always keep the promises they make to one another.
		3.	Members in the BlueShop virtual community would not knowingly do anything to disrupt the conversation.
		4.	Members in the BlueShop virtual community behave in a consistent manner.
		5.	Members in the BlueShop virtual community are truthful in dealing with one another.
	norm of reciprocity	1.	I know that other members in the BlueShop virtual community will help me, so it is only fair to help other members.
		2.	I believe that members in the BlueShop virtual community would help me if I need it.
	identification	1.	I feel a sense of belonging towards the BlueShop virtual community.
		2.	I have the feeling of togetherness or closeness in the BlueShop virtual community.
		3.	I have a strong positive feeling toward the BlueShop virtual community.
		4.	I am proud to be a member of the BlueShop community.
cognitive dimension	shared language	1.	The members in the BlueShop virtual community use common terms or jargons.
		2.	Members in the BlueShop virtual community use understandable communication pattern during the discussion.
		3.	Members in the BlueShop virtual community use understandable narrative forms to post messages or articles.
	shared vision	1.	Members in the BlueShop virtual community share the vision of helping others solve their professional problems.
		2.	Members in the BlueShop virtual community share the same goal of learning from each other.
		3.	Members in the BlueShop virtual community share the same value that helping others is pleasant.

In the next stage of measurement development, Yli-renko et al. (2001) selected items for the structural and relational dimensions from the existing one by Tsai and Goshal (1998), while developing new items for the cognitive one. In their paper, the structural dimension was termed as social interaction, while the relational one as relationship quality. Furthermore, the cognitive dimension was defined as customer network ties [45]. Wasko and Faraj (2005) proposed a self-rating scale for the cognitive dimension and applied the technique of Tsai and Ghoshal (1998) for the operationalization of the structural dimension. They also defined two subscales for the relational one (i.e., commitment and reciprocity), adapting previously operationalized scales from past literature [46–48]. These approaches were synthesized and validated in the aforementioned study by Chiu et al. (2006), which set a virtual, professional, IT-related community (i.e., BlueShop) in Taiwan as the subject of the analysis [43].

Although previous research indicates that the two previously discussed viewpoints constitute the majority of the empirical measurement approaches in terms of the perceived individual social capital on social media, the paper at hand intends to explore unique, emerging measurement techniques as well, to offer a broad and detailed scope for future studies. The present article aims to scope out the practical characteristics of the empirical studies evaluating social capital constructs on social media, therefore, measuring individual social capital. Based on the previous studies, social capital measurement techniques will be evaluated through a scoping review of 80 published studies to determine the measurement approaches used in past research. Papers employing bridging, bonding, and/or maintained social capital will be explored, followed by those of the multidimensional view, along with a discussion of unique social capital measurement approaches. The goal of the present paper is to: (i) span a broad and detailed scope, (ii) evaluate these techniques, and (iii) identify possible similarities or differences, to provide a more transparent view about the state of this research area and its possible empirical performance and explanatory power.

3. Search and Filtering Method

The scoping review methodology [49,50] was applied to map the current state of the scientific knowledge and identify possibly existing research gaps. A scoping review is appropriate here, as it provides an opportunity for a broader research question and the avoidance of bias-assessment.

A multi-keyword search was employed in ProQuest and Google Scholar (i.e., "social capital" AND ("social media" OR "social network" OR "SNS" OR "SM")). The collection of scientific literature followed a funnel approach [51]. Only peer-reviewed articles, peer-reviewed conference proceedings and peer-reviewed book chapters were included into the search criteria. The search process identified 2478 records.

Four manually performed filtering steps were performed on the 2478 records: (a) abstracts and reference lists were checked (139 records remaining), (b) quantitative studies were kept (65 records remaining), (c) studies that did not measure social capital were eliminated (53 records remaining), and (d) the citations of the remaining 53 papers were reviewed backwards and forwards. After the four steps were complete, the final set of $n = 80$ records remained.

Additional inclusion criteria for the final set of publications were as follows. The manuscript has to: (a) appear in a peer-reviewed article or conference proceedings or book chapter, (b) be written in the English language, (c) set individual, perceived social capital as the focus, (d) investigate this phenomenon on one or more social media platforms, and (e) empirically measure the perceived social capital in a quantitative manner.

The categorization for comparison and coding were performed using tables in Excel, involving the application of the cultural view, the multidimensional one, or a unique approach. The elements of the operationalized constructs were collected for evaluation based on consistency, joined with the collection of authors, whom the analyzed publications refer to in this regard. Furthermore, the dimension names were collected with attention to papers empirically investigating the multidimensional view of social capital, or using unique measurement approaches. This process involved the authors and two additional, independent reviewers stemming from the respective scientific areas of research.

4. Results

The analysis is based on the observations and trends extracted from the systematically collected literature. As previously indicated, two distinct operationalization techniques emerged from the analyzed $n = 80$ records: the majority, (i.e., two thirds (66%; 53 items)) of the analyzed publications investigated bridging, bonding, and/or maintained social capital constructs. These studies followed the theoretical considerations of Putnam (2000). The multidimensional view was explored by one-fifth (18%; 15 items) of the papers. These two viewpoints represent a contrast in terms of polarity, as they did not intersect regarding their direction of operationalization; however, none of the analyzed articles empirically compared these two concepts.

Bridging social capital was present in all empirical studies that operationalized the social capital constructs according to the cultural view, with the exception of one manuscript. Bonding capital appeared in almost all studies apart from six, thereby indicating its importance. Merely six studies quantified maintained social capital. The description of individual measurement items was explored, based on its cruciality for future replication possibilities. The review concluded that replication was not possible in 18% (nine items) of the studies interpreting bridging, bonding, and/or maintained social capital measurement on social media, as a lack of a measurement item description.

Through the analysis of individual measurement items, it became evident that there is a considerable diversity in terms of how many, and what kind of items the studies employed. Among the underlying reasons for difference is that the performed principal component analyses (PCA) and confirmatory factor analyses (CFA) delivered different results in individual studies, resulting in the exclusion of at least one or more measurement item. The measurement consistency in the cultural view of social capital was clearly visible in other terms, in the cases of all three measurement constructs (i.e., bridging, bonding, and maintained); however, considerable heterogeneity was found in the operationalization techniques of the multidimensional view.

Table 3 offers a summary of the measurement constructs of each social capital dimension according to the multidimensional view in the 15 analyzed papers, with the exception of Chiu and colleagues (2006).

As Table 3 illustrates, there are distinct differences in terms of sources for measurement operationalization and the construct names for all three dimensions. More specifically, Chiu and colleagues (2006) analyzed the structural dimension by employing one construct (social interaction ties). However, Table 3 shows a variety of construct names (e.g., social networking, instrumental network ties, expressive network ties) in this regard, combined with the diverse operationalization techniques. This trend is visible in terms of the relational and cognitive dimensions as well. It is, however, necessary to note that all studies analyzed in this view offered clear sources in applied measurement, combined with the availability of the measurement items, which can greatly enhance the possibility of replication and the ability of results generalization in a cumulative manner.

The majority of the sampled records employed either the cultural or the multidimensional view of social capital, along with their matching measurement techniques. Unique approaches are summarized in Table 4.

Table 4 reveals a high degree of consistency in the wording for social capital; however, quite distinct differences regarding operationalization techniques are observable as well. While all studies mentioned in Table 4 aimed to analyze the same theoretical concept (i.e., social capital), with a clear majority evaluating bridging, bonding and maintained constructs, the previously mentioned heterogeneity in operationalization discussed in the multidimensional view, extended with these unique approaches, further indicates that there is no particular measurement in this social capital view, which can be considered as common starting point. Quite the contrary, these results address the uncertainty regarding the construct measurement of social capital. Albeit the hypotheses aiming to find relationships with various constructs and social capital itself, were verified in the individual papers, they depicted these results through plentiful operationalization techniques.

Table 3. Evaluation of the 15 items (19%) of the final set of publications operationalizing social capital measurement constructs according to the multidimensional view. Abbreviations used: struct. = structural, dim. = dimension, constr. = construct, meas. = measurement(s), rel. = relational, cogn. = cognitive.

Study	Struct. Dim. Constr. Name(s)	Struct. Dim. Constr. Meas.	Rel. Dim. Constr. Name(s)	Rel. Dim. Construct Meas.	Cogn. Dim. Constr. Name(s)	Cogn. Dim. Constr. Meas.	Meas. Items Present in the Article (yes/no)
[52]	structural capital	[53,54]	relational capital	[45]	cognitive capital	[53,54]	yes
[55]	-	-	relational capital	[45,53]	-	-	yes
[56]	social networking	[57]	trust	[58]	shared language	[36]	yes
[59]	-	-	trust	[58]	-	-	yes
[60]	network ties	[36]	trust	[58]	shared vision	[43]	yes
[61]	social interaction	[43,45]	trust; identification; reciprocity	[43,48,62]	shared language	[43]	yes
[63]	-	-	social capital identification	[43,64]	-	-	yes
[65]	-	-	social capital identification	[43,64]	-	-	yes
[66]	instrumental network ties; expressive network ties	own items developed	identification; trust in online community; norms of cooperation	own items developed	shared language	[43]	no
[67]	social networking	[57]	trust	[58]	shared language	own items developed	no
[68]	-	-	trust	[59]	-	-	yes
[69]	social interaction ties	[43]	trust	[43]	shared vision	[43]	yes
[70]	social interaction ties	[43]	social trust	[57]	shared values	[71]	yes
[72]	structural capital	[43,57]	relational capital	[43,57]	cognitive capital	[43,57]	yes

Table 4. List and basis of comparison in the cases of 12 publications employing unique measurement approaches.

Study	Name of Social Capital	Measurement of Social Capital
[73]	social capital affinity	5 items partially adapted from the bridging measurement scales of [1,27].
[74]	social capital affinity	5 items partially adapted from the bridging measurement scales of [1,27].
[75]	social capital	users' view count on their individual videos and users' subscriber count on their channel
[76]	Karma	measured as the karma rankings of Slashdot users (see [76] for further details)
[77]	individual social capital	measured as the number of readers (equation for social capital is based on [78])
[79]	individual social capital	6-item index developed for social life features that represent effective collective action and pursuing shared objectives in the participants' communities
[80]	social capital	measured as physical (geographical) distance and strength of friendship (number of comments between two friends)
[81]	reciprocal tagging activity as the formation of bridging and bonding social capital	reciprocal actions in the form of liking/commenting a tagging activity or sharing the specific post the users were tagged in

Table 4. *Cont.*

Study	Name of Social Capital	Measurement of Social Capital
[82]	social capital	social capital as the degree of social participation (connectivity) + social support (content generation engagement). Connectivity = number of friends, number of community memberships, number of followers; social support = number of posts written, number of comments made, number of comments received
[83]	social capital	own measurement developed for civic engagement (1 item), interpersonal trust (2 items), political knowledge (6 items)
[84]	social capital	life satisfaction = adapted from the Satisfaction with Life Scale (developed by [85]); social trust = adapted and modified version of [86]; civic and political participation = a reduced form of the Index of Civic and Political Engagement developed by CIRCLE [87]
[88]	social capital	political participation = 6 items adapted from the National Election Studies; civic participation = 5 items developed; confidence in government = 3 items developed

5. Discussion and Conclusions

The present paper aimed to discover and evaluate prior empirical social capital research conducted in the realm of social media. The primary objective of the study was to tap into measurement operationalization techniques used for evaluating social capital, concentrating on cultural and multidimensional view approaches, and offering an extension into unique measurement approaches. Our analysis involved several tasks to provide a more transparent view about the state of the preceived individual social capital measurement on social media, and its possible empirical performance and explanatory power: (i) span a broad and detailed scope, (ii) evaluate the techniques, and (iii) identify possible similarities or differences. The paper intended to contribute to approaches, such as the meta-analytical approach in Liu et al. (2016), who observed the relationship of bridging and bonding social capital with global social media use and site activity. Such contributions can offer an opportunity of comparison and jointly reveal effect sizes of multiple records to answer the core question, whether the interpreted effects are existing, statistically significant, or the results of selective reporting [77–80].

From the viewpoint of interdisciplinary research, it seems necessary to discuss operationalization technique consistency and offer a synthesis to highlight that the possibility of future meta-analyses is strongly dependent on the comparability and coherence of measurement techniques to maintain validity in effect size measurement and the avoidance of system-inherent bias.

By means of a scoping review, the present study assessed 80 articles to evaluate the standing of social capital research on social media, concentrating on their operationalization techniques. While there is a general observable trend regarding the interpretation of individual measurement items and constructs, studies in the multidimensional view depicted great heterogeneity in terms of operationalization and proposed measurement techniques, which indicates challenging conditions for future meta-analytical approaches in this domain. On the other hand, studies employing the cultural view of social capital, along with the validated measurement techniques proposed by Williams (2006) or Ellison et al. (2007), show a high degree of consistency. It should be noted, however, that there is heterogeneity in the individual studies in terms of employed items from these scales, based on the results of the performed PCA and CFA analyses, resulting in possible item drops. Furthermore, unique social capital measurement techniques on social media are also present in this research domain, enhancing the complexity of a possible, empirical synthesis.

Social media platforms offer to fulfill the human drive to belong and have an exponentially growing user-base. The underlying motivations for the usage of such platforms, along with the expected and perceived benefits as a result of being present and active on them, are especially crucial to better understanding human behavior.

The present article aimed to provide a detailed view into the individual, perceived social capital research on social media, and limited itself for the discussion to articles exploring this phenomenon on at least one SM platform. However, as empirical social capital measurement is present in a plethora of further research fields in both an online and offline context, while investigated not merely in a perceived notion, nor solely on an individual level.

Social capital, which seems to be attached to a diverse set of behavioral phenomena [18], can be considered as one of such phenomena; therefore, its analysis, and possible synthesis is an ever pressing issue, since the concept of social capital is indeed a "buzzword" in science [89]. The wide array of measurement approaches discussed in this article, however, raise questions about the measurement: do they measure the same concept, or, as the opposite extreme, maybe none of them do.

The importance of social capital research on social media has possible individual benefits in terms of student learning outcomes, based on the discussed benefits of weak ties as an example. Further benefits include a diverse set of research areas, including the challenge of the cultural barriers for women's economic independence and autonomy [90], highlighting the importance of these investigations aiming to reduce inequalities. This cruciality also manifests itself in labor market studies, wherein individual social capital can be considered as an enabler for successful labor market integration [91]. It also manifests itself in healthcare research, since online conversations can possibly strengthen patient–caregiver connections, leading to successful online health communities, and ultimately, effective policy interventions [92]. Albeit, these examples are far from reaching comprehensivity, they do indicate the relevance of both the existence of social media for the benefits of humanity, and the diversity of areas on which social media can possibly provide benefits for individuals through enabling social capital.

Machine-learning based methods can further enhance the results of such empirical investigations, (e.g., sentiment analysis [93–97]), which could be employed as an extension to reveal the underlying sentiment in student communication present on forums, and class discussion boards. The usage of big data in data sciences, especially in the research area of digital marketing, indicate the crucial importance of such investigations, involving numerous industrial areas, detailed recently by Saura (2020). While companies aim to leverage from such methods, from which the author distinguished nine individual core topics [98], highlighting social media listening as well, the empirical research of individual, perceived social capital might offer crucial insights for corporations aiming to achieve effective digital marketing strategies. This implication is also supported by the relevant publications on the importance of social media marketing, wherein electronic word of mouth (e-WOM) is facilitated by user-generated content, which empowers customers with the ability of sharing their experiences about brands, products, or firms, in which trust plays a key role. [99] Trust is an essential part of the perceived, individual social capital according to the presently discussed views of individual social capital.

It is recommended that future research determines, in detail, how and in what manner, levels of individual online social capital on social media can possibly enable corporational profit enhancements through the mediating role of electronic word of mouth, possibly leading to more refined customer relationship management, accompanied with a positive brand perception.

Author Contributions: Conceptualization, F.P. and C.S.; data curation, F.P.; investigation, F.P.; methodology, F.P.; resources, F.P.; software, F.P.; supervision, C.S.; writing—original draft, F.P.; writing—review & editing, C.S. All authors have read and agreed to the published version of the manuscript.

Funding: Open Access Funding by the University of Vienna.

References

1. Ellison, N.B.; Steinfield, C.; Lampe, C. The benefits of facebook "friends:" Social capital and college students' use of online social network sites. *J. Comput. Commun.* **2007**, *12*, 1143–1168. [CrossRef]

2. Boyd, D.M.; Ellison, N.B. Social Network Sites: Definition, History, and Scholarship. *J. Comput. Commun.* **2007**, *13*, 210–230. [CrossRef]

3. VanMeter, R.A.; Grisaffe, D.B.; Chonko, L.B. Of "Likes" and "Pins": The Effects of Consumers' Attachment to Social Media. *J. Interact. Mark.* **2015**, *32*, 70–88. [CrossRef]

4. Kaplan, A.M.; Haenlein, M. Users of the world, unite! The challenges and opportunities of Social Media. *Bus. Horiz.* **2010**, *53*, 59–68. [CrossRef]

5. Jang, J.; Dworkin, J. Does social network site use matter for mothers? Implications for bonding and bridging capital. *Comput. Hum. Behav.* **2014**, *35*, 489–495. [CrossRef]

6. Baumeister, R.F.; Leary, M.R. The need to belong: Desire for interpersonal attachments as a fundamental human motivation. *Psychol. Bull.* **1995**, *117*, 497–529. [CrossRef]

7. Kim, Y.; Wang, Y.; Oh, J. Digital Media Use and Social Engagement: How Social Media and Smartphone Use Influence Social Activities of College Students. *Cyberpsychol. Behav. Soc. Netw.* **2016**, *19*, 264–269. [CrossRef]

8. Maslow, A.H. A Dynamic Theory of Human Motivation. In *Understanding Human Motivation*; Howard Allen Publishers: Cleveland, OH, USA, 1943; pp. 26–47.

9. Goffman, E. *The Presentation of Self in Everyday Life*; University of Edinburgh, Social Sciences Research Centre: Edinburgh, UK, 1956.

10. Winter, S.; Neubaum, G.; Eimler, S.C.; Gordon, V.; Theil, J.; Herrmann, J.; Meinert, J.; Krämer, N.C. Another brick in the Facebook wall—How personality traits relate to the content of status updates. *Comput. Hum. Behav.* **2014**, *34*, 194–202. [CrossRef]

11. Lin, N. *Social Capital: A Theory of Social Structure and Action*; Cambridge University Press: London, UK, 2001.

12. Wang, P.; Xie, X.; Wang, X.; Wang, X.; Zhao, F.; Chu, X.; Nie, J.; Lei, L. The need to belong and adolescent authentic self-presentation on SNSs: A moderated mediation model involving FoMO and perceived social support. *Pers. Individ. Dif.* **2018**, *128*, 133–138. [CrossRef]

13. Coffé, H.; Geys, B. Toward an empirical characterization of bridging and bonding social capital. *Nonprofit Volunt. Sect. Q.* **2007**, *36*, 121–139. [CrossRef]

14. Patulny, R.V.; Lind Haase Svendsen, G. Exploring the social capital grid: Bonding, bridging, qualitative, quantitative. *Int. J. Sociol. Soc. Policy* **2007**, *27*, 32–51. [CrossRef]

15. Granovetter, M. The Strength of Weak Ties. *Am. J. Sociol.* **1973**, *78*, 1360–1380. [CrossRef]

16. Granovetter, M. The Strength of Weak Ties: A Network Theory Revisited. *Sociol. Theory* **1983**, *1*, 201. [CrossRef]

17. Krämer, N.; Rösner, L.; Eimler, S.; Winter, S.; Neubaum, G. Let the Weakest Link Go! Empirical Explorations on the Relative Importance of Weak and Strong Ties on Social Networking Sites. *Societies* **2014**, *4*, 785–809. [CrossRef]

18. Fine, B. *Theories of Social Capital: Researchers Behaving Badly*; Pluto Press: London, UK, 2010.

19. Durlauf, S.N.; Fafchamps, M. Social Capital. In *Handbook of Economic Growth*; Elsevier Masson SAS: Issy-les-Moulineaux, France, 2005; Volume 1, pp. 1639–1699. ISBN 9780444520432.

20. Johnston, G.; Percy-Smith, J. In search of social capital. *Policy Polit.* **2003**, *31*, 321–334. [CrossRef]

21. Knorringa, P.; Van Staveren, I. Beyond social capital: A critical approach. *Rev. Soc. Econ.* **2007**, *65*, 1–9. [CrossRef]

22. Portes, A. Social Capital: Its Origins and Applications in Modern Sociology. *Annu. Rev. Sociol.* **1998**, *24*, 1–24. [CrossRef]

23. Bourdieu, P. The Forms of Capital. In *Handbook of Theory and Research for the Sociology of Education*; Richardson, J.G., Ed.; Greenwood Press: New York, NY, USA, 1986.

24. Coleman, J.S. Social Capital in the Creation of Human Captial. *Am. J. Sociol.* **1988**, *94*, S95–S120. [CrossRef]

25. Fukuyama, F. Social capital, civil society and development. *Third World Q.* **2001**, *22*, 7–20. [CrossRef]

26. Putnam, R.D. *Bowling Alone: The Collapse and Revival of American Community*; Simon and Schuster: New York, NY, USA, 2000.

27. Williams, D. On and Off the 'Net: Scales for Social Capital in an Online Era. *J. Comput. Commun.* **2006**, *11*, 593–628. [CrossRef]

28. Kraut, R.; Kiesler, S.; Boneva, B.; Cummings, J.; Helgeson, V.; Crawford, A. Internet Paradox Revisited. *J. Soc. Issues* **2002**, *58*, 49–74. [CrossRef]

29. Howard, P.E.N.; Rainie, L.; Jones, S. Days and nights on the Internet: The impact of a diffusing technology. *Am. Behav. Sci.* **2001**, *45*, 383–404. [CrossRef]

30. Price, V.; Cappella, J.N. Online Deliberation and its Influence: The Electronic Dialogue Project in Campaign 2000. *Itsociety* **2002**, *1*, 303–329.

31. Hanifan, L.J. The Rural School Community Center. *Ann. Am. Acad. Pol. Soc. Sci.* **1916**, *67*, 130–138. [CrossRef]

32. Hofer, M.; Aubert, V. Perceived bridging and bonding social capital on Twitter: Differentiating between followers and followees. *Comput. Hum. Behav.* **2013**, *29*, 2134–2142. [CrossRef]

33. Putnam, R.D.; Goss, K.A. Introduction. In *Democracies in Flux: The Evolution of Social Capital in Contemporary Society*; Oxford University Press: New York, NY, USA, 2002; pp. 3–21. ISBN 9780195150896.

34. Fine, B. *Social Capital Versus Social Theory: Political Economy and Social Science at the turn of the Millennium*; Routledge: London, UK, 2001; ISBN1 0415241790. ISBN2 9780415241793. ISBN3 0415241804. ISBN4 9780415241809.

35. Lin, N. Building a Network Theory of Social Capital. In *Social Capital Theory and Research*; Lin, N., Cook, K., Burt, R.S., Eds.; Transaction Publishers: New Brunswick, NJ, USA, 2001; pp. 3–30.

36. Nahapiet, J.; Ghoshal, S. Social Capital, Intellectual Capital, and the Organizational Advantage. *Acad. Manag. Rev.* **1998**, *23*, 242. [CrossRef]

37. Wang, S.; DesRoches, C.T.; Sun, L.; Stennes, B.; Wilson, B.; Cornelis van Kooten, G. Linking forests and economic well-being: A four-quadrant approach. *Can. J. For. Res.* **2007**, *37*, 1821–1831. [CrossRef]

38. Hunter, M.L. "If You're Light You're Alright"—Light Skin Color as Social Capital for Women of Color. *Gend. Soc.* **2002**, *16*, 175–193. [CrossRef]

39. Wood, L.; Giles-Corti, B.; Bulsara, M. The pet connection: Pets as a conduit for social capital? *Soc. Sci. Med.* **2005**, *61*, 1159–1173. [CrossRef] [PubMed]

40. Domahidi, E. The Associations Between Online Media Use and Users' Perceived Social Resources: A Meta-Analysis. *J. Comput. Commun.* **2018**, *23*, 181–200. [CrossRef]

41. Liu, D.; Ainsworth, S.E.; Baumeister, R.F. A meta-analysis of social networking online and social capital. *Rev. Gen. Psychol.* **2016**, *20*, 369–391. [CrossRef]

42. Quan-Haase, A.; Wellman, B. How does the Internet affect social capital? In *Social Capital and Information Technology*; Huysman, M., Wulf, V., Eds.; MIT Press: Cambridge, MA, USA, 2004; pp. 113–132.

43. Chiu, C.M.; Hsu, M.H.; Wang, E.T.G. Understanding knowledge sharing in virtual communities: An integration of social capital and social cognitive theories. *Decis. Support Syst.* **2006**, *42*, 1872–1888. [CrossRef]

44. Tsai, W.; Ghoshal, S. Social Capital and Value Creation: The Role of Intrafirm Networks. *Acad. Manag. J.* **1998**, *41*, 464–476. [CrossRef]

45. Yli-Renko, H.; Autio, E.; Sapienza, H.J. Social capital, knowledge acquisition, and knowledge exploitation in young technology-based firms. *Strateg. Manag. J.* **2001**, *22*, 587–613. [CrossRef]

46. Constant, D.; Sproull, L.; Kiesler, S. The Kindness of Strangers: The Usefulness of Electronic Weak Ties for Technical Advice. *Organ. Sci.* **1996**, *7*, 119–135. [CrossRef]

47. Mowday, R.T.; Steers, R.M.; Porter, L.W. The measurement of organizational commitment. *J. Vocat. Behav.* **1979**, *14*, 224–247. [CrossRef]

48. McLure Wasko, M.; Faraj, S. Why Should I Share? Examining Social Capital and Knowledge Contribution in Electronic Networks of Practice Why Should I Share? Examining Social Capital and Knowledge Contribution in electronic networks of practice1. *Source MIS Q.* **2005**, *29*, 35–57. [CrossRef]

49. Arksey, H.; O'Malley, L. Scoping studies: Towards a methodological framework. *Int. J. Soc. Res. Methodol.* **2005**, *8*, 19–32. [CrossRef]

50. Levac, D.; Colquhoun, H.; O'Brien, K.K. Scoping studies: Advancing the methodology. *Implement. Sci.* **2010**, *5*, 69. [CrossRef]

51. Webster, J.; Watson, R.T. Analyzing the Past to Prepare for the Future: Writing a Literature Review. *MIS Q.* **2002**, *26*, xiii–xxiii.

52. Bharati, P.; Zhang, W.; Chaudhury, A. Better knowledge with social media? Exploring the roles of social capital and organizational knowledge management. *J. Knowl. Manag.* **2015**, *19*, 456–475. [CrossRef]

53. Leana, C.R.; Pil, F.K. Social Capital and Organizational Performance: Evidence from Urban Public Schools. *Organ. Sci.* **2006**, *17*, 353–366. [CrossRef]

54. Teo; Wei; Benbasat Predicting Intention to Adopt Interorganizational Linkages: An Institutional Perspective. *MIS Q.* **2003**, *27*, 19. [CrossRef]

55. Bharati, P.; Chaudhury, A. Assimilation of Big Data Innovation: Investigating the Roles of IT, Social Media, and Relational Capital. *Inf. Syst. Front.* **2018**, *21*, 1357–1368. [CrossRef]

56. Cao, X.; Guo, X.; Liu, H.; Gu, J. The role of social media in supporting knowledge integration: A social capital analysis. *Inf. Syst. Front.* **2015**, *17*, 351–362. [CrossRef]

57. Chow, W.S.; Chan, L.S. Social network, social trust and shared goals in organizational knowledge sharing. *Inf. Manag.* **2008**, *45*, 458–465. [CrossRef]

58. Levin, D.Z.; Cross, R. The Strength of Weak Ties You Can Trust: The Mediating Role of Trust in Effective Knowledge Transfer. *Manag. Sci.* **2004**, *50*, 1477–1490. [CrossRef]

59. Cao, X.; Guo, X.; Vogel, D.R.; Liu, H.; Gu, J. Understanding the influence of social media in the workplace: An integration of media synchronicity and social capital theories. In Proceedings of the Annual Hawaii International Conference on System Sciences, Maui, HI, USA, 4–7 January 2012; pp. 3938–3947.

60. Cao, X.; Guo, X.; Vogel, D.; Zhang, X. Exploring the influence of social media on employee work performance. *Internet Res.* **2016**, *26*, 529–545. [CrossRef]

61. Chang, H.H.; Chuang, S.S. Social capital and individual motivations on knowledge sharing: Participant involvement as a moderator. *Inf. Manag.* **2011**, *48*, 9–18. [CrossRef]

62. Ridings, C.M.; Gefen, D.; Arinze, B. Some antecedents and effects of trust in virtual communities. *J. Strateg. Inf. Syst.* **2002**, *11*, 271–295. [CrossRef]

63. Huang, R.; Kim, H.J.; Kim, J. Social capital in QQ China: Impacts on virtual engagement of information seeking, interaction sharing, knowledge creating, and purchasing intention. *J. Mark. Manag.* **2013**, *29*, 292–316. [CrossRef]

64. Dholakia, U.M.; Bagozzi, R.P.; Pearo, L.K. A social influence model of consumer participation in network- and small-group-based virtual communities. *Int. J. Res. Mark.* **2004**, *21*, 241–263. [CrossRef]

65. Kim, H.; Kim, J.; Huang, R. Social Capital in the Chinese Virtual Community: Impacts on the Social Shopping Model for Social Media. *Glob. Econ. Rev.* **2014**, *43*, 3–24. [CrossRef]

66. Law, S.P.-M.; Chang, M.K. Fostering Knowledge Exchange in Online Communities: A Social Capital Building Approach. In Proceedings of the ICIS 2008, Paris, France, 14–17 December 2008; pp. 1–22.

67. Mei, Y. Contribution of Social Media Use at Work to Social Capital and Knowledge Sharing: A Comparison between Chinese and Thai Employees. *Int. J. Appl. Comput. Technol. Inf. Syst.* **2016**, *6*, 21–27.

68. Offong, G.O.; Costello, J. Enterprise social media impact on human resource practices. *Evid. Based HRM A Glob. Forum Empir. Scholarsh.* **2017**, *5*, 328–343. [CrossRef]

69. Okazaki, S.; Andreu, L.; Campo, S. Knowledge Sharing Among Tourists via Social Media: A Comparison Between Facebook and TripAdvisor. *Int. J. Tour. Res.* **2017**, *19*, 107–119. [CrossRef]

70. Son, J.E.; Lee, S.H.; Cho, E.Y.; Kim, H.W. Examining online citizenship behaviours in social network sites: A social capital perspective. *Behav. Inf. Technol.* **2016**, *35*, 730–747. [CrossRef]

71. Zhao, L.; Lu, Y. Enhancing perceived interactivity through network externalities: An empirical study on micro-blogging service satisfaction and continuance intention. *Decis. Support Syst.* **2012**, *53*, 825–834. [CrossRef]

72. Wu, Y.-L.; Li, E.Y.; Chang, W.-L. Nurturing user creative performance in social media networks. *Internet Res.* **2016**, *26*, 869–900. [CrossRef]

73. Barker, V.; Dozier, D.M.; Weiss, A.S.; Borden, D.L. Facebook "friends": Effects of social networking site intensity, social capital affinity, and flow on reported knowledge-gain. *J. Soc. Media Soc.* **2013**, *2*, 76–97.

74. Barker, V.; Dozier, D.M.; Weiss, A.S.; Borden, D.L. Harnessing peer potency: Predicting positive outcomes from social capital affinity and online engagement with participatory websites. *New Media Soc.* **2015**, *17*, 1603–1623. [CrossRef]

75. Feroz Khan, G.; Vong, S. Virality over YouTube: An empirical analysis. *Internet Res.* **2014**, *24*, 629–647. [CrossRef]

76. Ganley, D.; Lampe, C. The ties that bind: Social network principles in online communities. *Decis. Support Syst.* **2009**, *47*, 266–274. [CrossRef]

77. Gaudeul, A.; Giannetti, C. The role of reciprocation in social network formation, with an application to LiveJournal. *Soc. Netw.* **2013**, *35*, 317–330. [CrossRef]

78. Glaeser, E.; Laibson, D.; Sacerdote, B. An Economic Approach to Social Capital. *Econ. J.* **2002**, *112*, F437–F458. [CrossRef]

79. Gil de Zúñiga, H.; Jung, N.; Valenzuela, S. Social Media Use for News and Individuals' Social Capital, Civic Engagement and Political Participation. *J. Comput. Commun.* **2012**, *17*, 319–336. [CrossRef]

80. Gilbert, E.; Karahalios, K.; Sandvig, C. The network in the garden: Designing social media for rural life. *Am. Behav. Sci.* **2010**, *53*, 1367–1388. [CrossRef]

81. Ha, T.; Han, S.; Lee, S.; Kim, J.H. Reciprocal nature of social capital in Facebook: An analysis of tagging activity. *Online Inf. Rev.* **2017**, *41*, 826–839. [CrossRef]

82. Nguyen, T.; Dao, B.; Phung, D.; Venkatesh, S.; Berk, M. Online Social Capital: Mood, Topical and Psycholinguistic Analysis. In Proceedings of the 7th AAAI International Conference on Weblogs and Social Media, Cambridge, MA, USA, 8–11 July 2013; pp. 449–456.

83. Pasek, J.; More, E.; Romer, D. Realizing the social internet? Online social networking meets offline civic engagement. *J. Inf. Technol. Polit.* **2009**, *6*, 197–215. [CrossRef]

84. Valenzuela, S.; Park, N.; Kee, K.F. Is There social capital in a social network site?: Facebook use and college student's life satisfaction, trust, and participation1. *J. Comput. Commun.* **2009**, *14*, 875–901. [CrossRef]

85. Diener, E.; Emmons, R.A.; Larsen, R.J.; Griffin, S. The Satisfaction with Life Scale. *J. Pers. Assess.* **1985**, *49*, 71–75. [CrossRef]

86. Rosenberg, M. Misanthropy and Political Ideology. *Am. Sociol. Rev.* **1956**, *21*, 690–695. [CrossRef]

87. Andolina, M.; Keeter, S.; Zukin, C.; Jenkins, K. A Guide to the Index of Civic and Political Engagement. Available online: https://www.researchgate.net/publication/267399505_A_guide_to_the_index_of_civic_and_political_engagement (accessed on 3 November 2020).

88. Zhang, W.; Johnson, T.J.; Seltzer, T.; Bichard, S.L. The revolution will be networked: The influence of social networking sites on political attitudes and behavior. *Soc. Sci. Comput. Rev.* **2010**, *28*, 75–92. [CrossRef]

89. Solow, R.M. But Verify. *New Repub.* **1995**, *213*, 36–39.

90. Ali Aksar, I.; Danaee, M.; Maqsood, H.; Firdaus, A. Women's social media needs and online social capital: Bonding and bridging social capital in Pakistan. *J. Hum. Behav. Soc. Environ.* **2020**, *00*, 1–24. [CrossRef]

91. Gericke, D.; Burmeister, A.; Löwe, J.; Deller, J.; Pundt, L. How do refugees use their social capital for successful labor market integration? An exploratory analysis in Germany. *J. Vocat. Behav.* **2018**, *105*, 46–61. [CrossRef]

92. Panzarasa, P.; Griffiths, C.J.; Sastry, N.; de Simoni, A. Social medical capital: How patients and caregivers can benefit from online social interactions. *J. Med. Internet Res.* **2020**, *22*, 1–4. [CrossRef]

93. Poecze, F.; Ebster, C.; Strauss, C. Social media metrics and sentiment analysis to evaluate the effectiveness of social media posts. *Procedia Comput. Sci.* **2018**, *130*, 660–666. [CrossRef]

94. Poecze, F.; Ebster, C.; Strauss, C. Let's play on Facebook: Using sentiment analysis and social media metrics to measure the success of YouTube gamers' post types. *Pers. Ubiquitous Comput.* **2019**. [CrossRef]

95. Hew, K.F.; Hu, X.; Qiao, C.; Tang, Y. What predicts student satisfaction with MOOCs: A gradient boosting trees supervised machine learning and sentiment analysis approach. *Comput. Educ.* **2020**, *145*, 103724. [CrossRef]

96. Troisi, O.; Grimaldi, M.; Loia, F.; Maione, G. Big data and sentiment analysis to highlight decision behaviours: A case study for student population. *Behav. Inf. Technol.* **2018**, *37*, 1111–1128. [CrossRef]

97. Kumar, A.; Jain, R. Sentiment analysis and Feedback Evaluation. In Proceedings of the 2015 IEEE 3rd International Conference on MOOCs, Innovation and Technology in Education (MITE), Amritsar, India, 1–2 October 2015; pp. 433–436.

98.   Saura, J.R. Using Data Sciences in Digital Marketing: Framework, methods, and performance metrics. *J. Innov. Knowl.* **2020**. [CrossRef]
99.   Alalwan, A.A.; Rana, N.P.; Dwivedi, Y.K.; Algharabat, R. Social media in marketing: A review and analysis of the existing literature. *Telemat. Inform.* **2017**, *34*, 1177–1190. [CrossRef]

Publisher's Note: MDPI stays neutral with regard to jurisdictional claims in published maps and institutional affiliations.

Article

Towards Context-Aware Opinion Summarization for Monitoring Social Impact of News

Alejandro Ramón-Hernández [1], Alfredo Simón-Cuevas [2,*], María Matilde García Lorenzo [1], Leticia Arco [3] and Jesús Serrano-Guerrero [4]

[1] Centro de Investigaciones de la Informática, Universidad Central "Marta Abreu" de Las Villas, Villa Clara 54830, Cuba; aramon@uclv.cu (A.R.-H.); mmgarcia@uclv.edu.cu (M.M.G.L.)

[2] Facultad de Ingeniería Informática, Universidad Tecnológica de La Habana "José Antonio Echeverría", La Habana 11500, Cuba

[3] Computer Science Department, Vrije Universiteit Brussel, 1050 Brussels, Belgium; larcogar@vub.be

[4] Department of Technologies and Information Systems, University College of Computer Science, University of Castilla-La Mancha, 13071 Ciudad Real, Spain; Jesus.Serrano@uclm.es

* Correspondence: asimon@ceis.cujae.edu.cu; Tel.: +53-7-2607242

Received: 10 October 2020; Accepted: 13 November 2020; Published: 18 November 2020

Abstract: Opinion mining and summarization of the increasing user-generated content on different digital platforms (e.g., news platforms) are playing significant roles in the success of government programs and initiatives in digital governance, from extracting and analyzing citizen's sentiments for decision-making. Opinion mining provides the sentiment from contents, whereas summarization aims to condense the most relevant information. However, most of the reported opinion summarization methods are conceived to obtain generic summaries, and the context that originates the opinions (e.g., the news) has not usually been considered. In this paper, we present a context-aware opinion summarization model for monitoring the generated opinions from news. In this approach, the topic modeling and the news content are combined to determine the "importance" of opinionated sentences. The effectiveness of different developed settings of our model was evaluated through several experiments carried out over Spanish news and opinions collected from a real news platform. The obtained results show that our model can generate opinion summaries focused on essential aspects of the news, as well as cover the main topics in the opinionated texts well. The integration of term clustering, word embeddings, and the similarity-based sentence to-news scoring turned out the more promising and effective setting of our model.

Keywords: opinion mining; opinion summarization; topic modeling; semantic similarity measures; word embeddings

---

1. Introduction

The globalization of the use of the Internet and the development of technologies such as Cloud Computing, Internet of Things, social networks, Mobile Computing, and others has favored the increase of user-generated content on the web. Nowadays, a surprisingly high quantity of news, messages, and reviews of products or services are generated in online social media, news portals, e-commerce sites, etc. The data and information produced by users have proven useful in many domains (e.g., marketing studies, business intelligence, health, governance, and others) [1]. The processing of user-generated content on digital platforms (e.g., news platforms) is playing significant roles in the success of government programs and initiatives in digital governance, from extracting and analyzing citizens' sentiments for decision-making [2]. Several efforts have been dedicated to deal with extracting knowledge and efficient processing of this unstructured information produced by users [3], resulted in increasing research interest in tasks within Natural Language Processing (NLP) such as sentiment analysis, also called opinion mining [4].

Opinion mining is the field of study that analyzes people's opinions, sentiments, appraisals, attitudes, and emotions towards entities and their attributes expressed in written texts [3]. Opinion mining (or sentiment analysis) is a broad area that includes many tasks, such as sentiment classification, aspect-based sentiment analysis, lexicon construction, opinion summarization, and others [5]. Opinion summarization is the task of automatically generating summaries for a set of opinions that are related to the same topic or specific target [6]. The aspect-based opinion summarization is one of the main approaches [7], but it would not be very appropriate in contexts where the opinions are not about products or services (e.g., opinions about news). Although summaries generated by several of the reported approaches are focused on specific topics [1,8,9], they are generally identified by looking only at the content in opinionated texts, whereas the context that originates the opinions (e.g., news) is not usually taken into account, being this a weakness. A comprehensive summary of the users' reactions concerning a news article can be crucial due to various reasons, such as (1) understanding the sensitivity/importance of the news, (2) obtaining insights about the diverse opinions of the readers regarding the news, and (3) understanding the key aspects that draw the interest of the readers [10]. On the other hand, to integrate both topic-opinion analysis and semantic information can yield satisfactory results in opinion summarization [1]. Nevertheless, the use of WordNet [11], as well as the deep-learning-based word embedding [12,13] (e.g., word2vec [14]) to represent and analyze the semantic of words when dealing with opinion summarization problems has been limited. Our work is addressed to the application of these models and resources to cope with opinion summarization challenges.

In this paper, a news-focused opinion summarization model is presented, which is conceived according to the conception of extractive and topic-based text summarization methods. Our model combines topic modeling, sentiment analysis, and the news-focused relevance scoring in seven phases: preprocessing, topic detection, sentiment scoring, topic-sentence mapping, topic contextualization, sentence ranking, and summary construction. The integration of these techniques allows us to deal with the problem in which the relevance focus not only comes from the texts of the opinions, but also comes from the news articles as the context that originates them. Semantic analysis is included in several phases, to improve text processing. The semantic characteristics of words are captured through the word2vec representation model [14] and from WordNet [11]. Besides, semantic similarity measures are used to assess the semantic relatedness between sentences-to-sentences and sentences-to-news.

The model was evaluated across two datasets containing Spanish news and opinions collected from a real digital news platform. The selected news and opinions are related to telecommunication services and the COVID-19 pandemic. The performance of our proposal was measured, using the Silhouette [15] and the Jensen–Shannon divergence (JSD) measures [16]. The first one is used to measure the quality of the clustering process, and then to estimate the prospective quality of the topic detection phase. The second one is used to measure the quality of the obtained summaries. Several experiments were carried out, to provide a deeper grounding for the contribution of our approach. Different settings of the proposed model were evaluated and compared, to analyze the behavior of the different techniques integrated into the model and to identify the best solution for the news-focused opinion summarization process. The analysis of the experimental results and obtained conclusions were substantiated through the well-known Wilcoxon's Statistics Test.

The rest of the paper is organized as follows: Section 2 summarizes the analysis of related works; Section 3 describes the proposed opinion summarization model; and Section 4 presents the datasets, metric description, and the experimental results and discussion. Conclusions and future work are pointed out in Section 5.

2. Related Works

Automatic text summarization is the task of producing a concise and fluent summary, condensing the most relevant and essential information contained in one or several textual documents, while preserving key information content and overall meaning of the information source [17]. Summarizing texts is still an active research field and needs further developments due to the huge data increase on the web [18] (e.g., user-generated content). These methods and techniques have been addressed for processing user-generated opinionated content on social networks and digital platforms, emerging as a new challenge [6]. Summaries can be automatically obtained through extractive (i.e., selecting the most important sentences from documents) or abstractive methods (i.e., generating new cohesive text that may not be present in the original information) [6,19]. Most of the opinion summarization models follow extractive methods [7,20]. Unlike traditional text summarization, the opinion-oriented summaries have to take into consideration the sentiment a person has towards a topic, product, place, or service [1]. Since a text summarization aims to generate a concise version of factual information, a sentiment summarization summarizes sentiments from a large number of reviewers or multiple reviews [21]. The opinion mining provides the sentiment associated with a document at different levels through the polarity detection task, whereas text summarization techniques identify the most relevant parts of one or more documents and build a coherent fragment of text (the summary) from them [1].

One of the main approaches to generate opinion summaries is the aspect-based opinion summarization [7,22], which summarizes opinions depending on different aspects or features (attributes or components) of an entity (objects, organizations, services, and products). In the context in which the aspects or features do not stand out, topic detection turning out critical for dismissing non-relevant sentences. However, achieving high effectiveness in this process constitutes a challenging task in contexts of the great diversity of opinions. Identifying topics is of great importance to determine regarding which issues users are giving their criteria [23], being one of the reasons that some opinion summarization approaches detect topics in their textual analysis [1,8,9,24,25]. Although the resulting summaries are generally focused on aspects or topics, they are mainly identified taking into account only the content of the opinionated texts and do not focus on specific information-context interests. Nevertheless, there are approaches where the relevance focus not only comes from the texts of the opinions, such as query-based opinion summarization, which aims to extract and summarize the opinionated sentences related to the user's query [6,26,27]. In these systems, classical summarization techniques are applied, and the context (query) is used as a relevant focus, to generate a coherent and useful summary for the user [28]. Other challenges are implicit in these opinion summarization methods, such as the following: how to retrieve query relevant sentences, how to cover the main topics in the opinionated text set, and how to balance these two requests [29]. Our proposal is addressed to a similar problem, where news articles are used as the relevant focus instead of users' queries, although few approaches dealing with this problem have been identified [10]. For instance, Chakraborty et al. reported a method of summarizing news article tweets that initially captures the diverse opinions from the tweets by creating a unique tweet similarity graph, followed by a community detection technique to identify the tweets representing these diverse opinions [10]. Representative keywords of the news articles are extracted to identify related tweets. The similarity scoring between news-tweets and a pair of tweets is based on the overlapping keywords (content similarity), and the word vectors' similarity (context similarity), respectively.

According to the results reported in Reference [1], integrating both topic-opinion analysis and semantic information can yield satisfactory results in opinion summarization. In this sense, for the analysis of opinions which are generally short texts, it is more useful to represent terms and to capture semantic information about them. Two fundamental approaches collect semantic characteristics of terms. One of them depends on the context, and the other one depends on the meaning. Latent Semantic Analysis (LSA) and Latent Dirichlet Allocation (LDA) are more commonly used methods for topic modeling in opinions and to capture the semantic information from the context, as reported in [1,8,24,25]. However, some researchers consider LDA- and LSA-based approaches to not proprerly model the aspects of the reviews made on the web [3]; instead, clustering text segment approaches have the advantage of keeping the document structure through segments, to capture the semantics of texts [30]. On the other hand, word embedding models [12] (e.g., word2vec [14], Glove [31], and FastText) have been less applied; only a few approaches have been identified [8,10]. A word embedding is a learned representation for text where words that have the same meaning have a similar representation. This kind of representation has been successful in extractive summarization [32]. WordNet [11] is the most commonly used technique for capturing and processing the semantic meaning of terms; however, it has not been so much when summarizing opinions. In this context, the use of WordNet is mainly limited to capture synonyms, and few approaches have been identified [26,33,34]. Nevertheless, the use of WordNet in our proposal goes further on.

3. News-Focused Opinion Summarization Model

The conception of the proposed model is based on the extractive and topic-based text summarization approach, where the relevance scoring of sentences not only requires processing the information content to be summarized (e.g., the set of opinions), but also requires to carry out an alignment process with external or contextual information of interest—in our case, news content. An overview of the proposed model is shown in Figure 1. The proposed model combines the topic modeling (phase 2) and the news content, to determine the "importance" of opinionated sentences; it also includes the sentiment analysis process (phase 3) to determine the polarity strength of sentences and avoid the inclusion of non-opinionated sentences in the automatic summary. The topic-sentence mapping (phase 4) and topic contextualization (phase 5) allow us to align the sentences to the corresponding identified opinion topics and to determine the most relevant topics concerning the news. The least relevant topics are discarded, following the sentence ranking (phase 6) and summary construction (phase 7) processes.

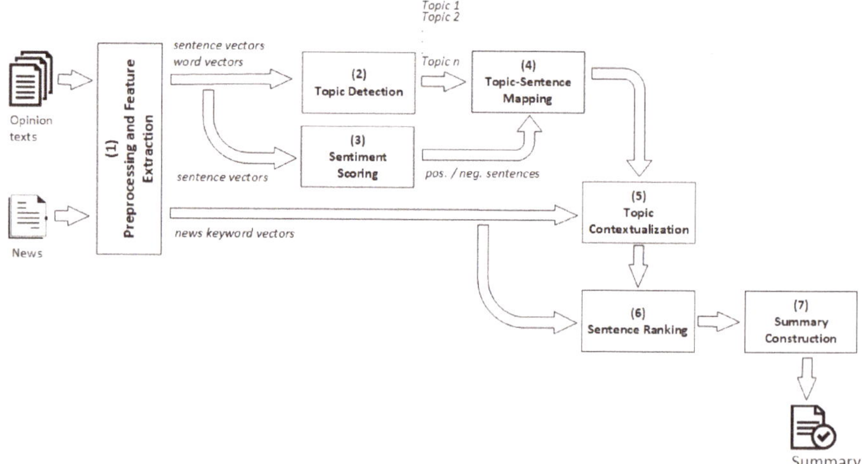

Figure 1. Workflow overview of the proposed model.

Several model settings and techniques were developed and evaluated, which are centered to address three important problems in the proposed model, such as (1) granularity in the topic modeling, (2) semantic processing of words and sentences, and (3) sentence relevance scoring. All of these developed alternatives are explained in the following subsections.

3.1. Preprocessing and Feature Extraction

In this phase, several Natural Language Processing tasks are performed for structuring the text (news and opinions) and extracting features, according to the preprocessing steps commonly reported in the opinion mining solutions [4]. Initially, the texts are split into sentences, and the tokenization task is applied to each sentence, for obtaining words or phrases. Some stop words, such as "la", "de", "y" and "o" (experiments were developed using Spanish text), are removed, considering that these words provide little useful information. Besides this, the lemmatization process of all words is carried out. Subsequently, the Part-of-Speech (POS) tagging is performed to determine the POS tag corresponding to each word belonging to sentences that make up opinions and news. The spaCy library of Python was used to support these tasks.

A crucial phase in opinion summarization is the feature-extraction phase, which simplifies the complexity of the involved tasks (e.g., topic modeling, sentiment classification, and semantic processing) by reducing the feature space. POS tags, such as adjective and noun, are quite helpful because the opinion words are usually adjectives and opinion targets (e.g., entities, aspects, or topics) are nouns or combinations of nouns [4]. Consequently, opinion features are constituted by noun phrases, adjectives, and adverbs. In the case of news texts, noun phrases play an important role as keywords in the content; therefore, they are used to construct the news keyword vector.

The vector space model was adopted for representing words and sentences (features). Two semantic representation approaches to reinforce the semantic processing were developed and evaluated, which are conceived through the use of (1) WordNet [11] and (2) word embeddings [12]. WordNet groups nouns, verbs, adjectives, and adverbs into sets of cognitive synonyms (*synsets*), each expressing a distinct concept meaning. Synsets are interlinked by means of conceptual–semantic and lexical relations. In the first case, the semantic characteristics of words are captured depending on their meaning. The feature vector is constructed with the *synset* of each word included in the sentence; in the case of ambiguous words (more than one *synset* in WordNet), the first *synset* that appears is selected. In the second case, the semantic characteristics of words are captured depending on their context. Word embedding vectors are obtained by applying the automatic learning model word2vec [14] on the sentences and news texts. Specifically, those vectors are generated by using the word2vec pre-trained model included in the es_core_news_md model of the spaCy library, which includes 300-dimensional vectors trained using FastText CBOW on Wikipedia and OSCAR (Common Crawl) containing 20 k unique words in Spanish.

3.2. Topic Detection

Topic detection is a way for monitoring and summarizing information generated from social sources, about which the participants discuss or argue or express their opinions. Therefore, identifying topics is of great importance to determine the relevant sentences of the opinion source to be included in the automatic summary. A topic can be analyzed and represented by considering different textual unit granularity, such as a group of terms, keywords, or sentences [30]. Term and sentence-based topic modeling approaches were applied and evaluated, adopting finally the first one in our proposal, as a consequence of the experimental results.

In our proposal, topic detection from all opinions is based on a clustering process, specifically of the terms extracted in the preprocessing task. In this sense, the cluster of terms represents the topics that have been boarded in the opinions. The objective of the clustering algorithms is to create groups that are coherent internally. In brief, cluster analysis groups data objects into clusters such that objects belonging to the same cluster are similar, while those belonging to different ones are dissimilar [35]. Both term and sentence clustering are carried out by applying a Hierarchical Agglomerative Clustering (HAC) algorithm [35]. HAC algorithm build hierarchies until obtaining a single cluster where all the objects are included. However, we need to obtain a certain quantity of groups of sentences that represent the topics boarded in the opinions. In this way, it is necessary to cut the hierarchy at some level for obtaining a partition. Although some variants to obtain a partition from a dendrogram are reported in Reference [35], we adopted the definition of a threshold to achieve a standard cut-point for the hierarchies, which allows us to compare the results of the similarity measures of the clusters with this threshold in the cluster-construction process. Thus, terms are clustered until their higher similarities are less than the specified threshold; otherwise, the clustering process will be stopped. To obtain the threshold value, the mean of the maximum values of the similarities among any pair of objects was considered.

Two semantic processing approaches for measuring the similarity between text units in the clustering process were evaluated: (1) WordNet and (2) word embedding based, with the last one being the most promising. The Wu and Palmer measure included in WordNet::Similarity [36] is applied for computing the similarity of terms where the WordNet-based semantic processing is applied. The cosine similarity measure is applied over the word embeddings based term representation. The similarity between the sentences S_1 and S_2 is determined by using the following sentence-to-sentence similarity function [37] expressed in Equation (1):

$$sem\_sim(S_1,\ S_2) = \frac{1}{2}\left(\frac{\sum_{w\in\{S_1\}}(maxSim(w,\ S_2) * idf(w))}{\sum_{w\in\{S_1\}} idf(w)} + \frac{\sum_{w\in\{S_2\}}(maxSim(w,\ S_1) * idf(w))}{\sum_{w\in\{S_2\}} idf(w)}\right) \quad (1)$$

In this function, given two sentences, S_1 and S_2, for each word (w) in S_1, it is identified the word w' in the sentence S_2 that has the highest semantic similarity $maxSim(w_i, S_2)$, according to one of the word-to-word similarity measures (in our proposal, Wu and Palmer or cosine measures).

3.3. Sentiment Scoring

Different from traditional extractive text summarization, whose fundamental goal is extracting "important" sentences from single or multi-documents according to some features, the opinion-oriented summaries have to take into consideration the sentiment a person has towards a topic, product, place, service, etc. Opinion mining provides the sentiment associated with a document at different levels and through the polarity detection task, whereas text summarization techniques identify the most relevant parts of a document and build from them a coherent fragment of text (the summary) [1].

In this step, the sentiment analysis processing is performed based on a lexicon-based method, using the SpanishSentiWordNet (Spanish adjustment of SentiWordNet [38]) to extract sentiment-related words in texts. The SpanishSentiWordNet [39] lexicon is the result of the automatic annotation of all *synsets* of Spanish WordNet, according to the notions of "positivity" and "negativity". In this process, each WordNet *synset* is associated with two numerical scores, which indicate degrees of positivity and negativity of the contained terms (noun, verb, adjective, and adverb) in the *synset* [39]. The sentences that do not include sentiment content, or that have lower sentiment scores than a threshold value, are filtered. Words with a positive or negative SpanishSentiWordNet score greater than 0.4 are considered when computing the sentiment scores. The polarity scoring of a sentence is calculated as shown in Equations (2) and (3) [30]:

$$PosSentenceScore(j) = \sum_{t_i\in Opinion(j)} PosValue(t_i) \quad (2)$$

$$NegSentenceScore(j) = \sum_{t_i\in Opinion(j)} NegValue(t_i) \quad (3)$$

where $PosValue(t_i)$ and $NegValue(t_i)$ are the polarity values in SpanishSentiWordNet of the identified sentiment word t_i in the opinion j. The opinion polarity is determined according to the highest obtained polarity scores. According to Reference [30], the sum operator reached better accuracy achieved in the experimental results between four compared classical compensatory operators. The topic polarity scores are measured by using the sum of the polarity scores *PosSentenceScore(S_j)* and *NegSentenceScore(S_j)* of each sentence S_j included in each cluster, according to Equations (4) and (5).

$$PosTopicScore(i) = \sum_{S_j \in Cluster(i)} PosSentenceScore\left(S_j\right) \tag{4}$$

$$NegTopicScore(i) = \sum_{S_j \in Cluster(i)} NegSentenceScore\left(S_j\right) \tag{5}$$

The highest obtained value of the cluster polarity score (*TopicScore(i)*) is used for determining which judgment (positives or negative) about the detected topics is the most representative in the processed opinion.

3.4. Topic-Sentence Mapping

Topic-based opinion-summarization systems, as our proposal, should be able not only to detect sentences that express a sentiment, but, more important, they should detect sentences that contain sentiment expressions towards the topic we are considering [1]. Once the opinion topics are identified and the sentences are classified as positive or negative, a mapping process between topics and sentences is performed. This process avoids the introduction of irrelevant sentences in the automatic summary. Mapping is carried out through computing the semantic similarity between the vocabulary that describes the topic and the sentences. For each sentence, Equation (1) is applied to compute sentences-to-topic similarity scores concerning all identified topics. Finally, the sentence is mapped onto the topic of the highest similarity score.

3.5. Topic Contextualization

Topic contextualization is one of the distinguishing tasks of our methodological proposal, concerning the generic opinion summarization systems that have been reported. In those systems, the generated summaries are generally focused on aspects or topics that are mainly identified while taking into account only the content of the opinionated texts. However, the purpose of our model is to provide automatic summaries focused on contexts of interest. In our model, these contexts are news articles, due the to fact they are the generators of the opinion comments.

In this phase, the news-based topic-ranking process is performed through computing the topic salience concerning the news content, obtaining a salience score for each topic. The topic salience is obtained by measuring the semantic similarity between the vocabulary associated with the topic and the news content. Topics with the lowest score (smaller or equal to a predefined threshold, which empirically was fixed in 0.5) are eliminated for the next steps of the summary construction process. This procedure means that the automatic summary will be built by extracting sentences from relevant topics of the news.

Similar to previous phases, Equation (1) and the conception for word-to-word semantic similarity are also applied. Topics are represented through term vectors, since the news is represented through the previously generated news feature vector. Formally, the salience score of a topic T_i for piece of news n_j is defined according to Equation (6). In the case of using sentence-based topic modeling (another developed and evaluated approach), topic salience is computed by averaging the semantic similarity between the sentence $S_k/S_k \in T_i$ and the news keyword vector, as shown in Equation (7).

$$salience\_score_1\left(T_i, n_j\right) = sem\_sim\left(T_i, n_j\right) \tag{6}$$

$$salience\_score_2\left(T_i, n_j\right) = \frac{\sum_{S_k \in T_i} sem\_sim\left(S_k, n_j\right)}{|T_i|} \tag{7}$$

3.6. Sentences Ranking

In this phase, the relevance assessment process applied to each opinionated sentence is carried out for generating the sentence ranking, according to a relevance score. Three approaches were developed and evaluated for measuring the relevance score:

1. Explanatoriness scoring [40]: In this approach, the ranking of sentences in opinions is based on their usefulness for helping users understand the reasons of sentiments (e.g., "explanatoriness"). It is one of the reported proposals in which the context is considered for determining the importance of the sentences. Kin et al. [40] proposed three heuristics for scoring explanatoriness of a sentence (i.e., length, popularity, and discriminativeness):

    *   *Sentence length*: A longer sentence is very likely to be more explanatory than than a shorter one, since a longer sentence, in general, conveys more information.
    *   *Popularity and representativeness*: A sentence is very likely to be more explanatory if it contains more terms that occur frequently in all sentences.
    *   *Discriminativeness relative to background*: A sentence containing more discriminative terms that can distinguish opinionated sentences from background information is more likely explanatory.

    In our proposal setting, for each sentence S_k, the clustered content by the contextualized topic to which the sentence S_k belongs is used as a reference for computing the *representativeness*. In addition, sentences from all opinions are used as background for computing the *discriminativeness*. It is important to point out that contextualized topics are the most important opinion topics for the news; therefore, this setting allows us to indirectly align the sentence relevance scoring process with the news context.

2. TextRank scoring [41]: TextRank is one of the most recognized standard and popular text summarization methods. This approach is conceived as a graph-based ranking model that is applied to an undirected graph extracted from natural language texts. In the graph, a sentence is represented as a vertex, and the "similarity" relation between two sentences determines the connexion (edge) between them. PageRank algorithm [42] is applied for computing the importance of a vertex (i.e., a sentence) within a graph.

3. Sentences-to-news scoring: This approach consists of computing the relevance score of each sentence S_k through measuring the semantic similarity between the sentence and the keyword vector of the news. For this purpose, Mihalcea et al. similarity function [37] (Equation (1)) is applied. Besides, two variants of the word-to-word semantic similarity are evaluated. Different from the explanatoriness scoring conception, this approach allows us to directly put the sentence-relevance scoring process in alignment with the news context, with the independence of the topic to the one belongs.

3.7. Summary Construction

Once the relevance of the sentences is computed in the previous phase, the summary-construction process is carried out by selecting the N opinionated sentences with a higher relevance score from each contextualized relevant topic. The N value depends on the predefined compression rate (summary size). However, we set $N = 3$ when evaluating our proposal.

4. Experimental Results

4.1. Description of Datasets

To evaluate the effectiveness of our proposed model, two datasets with real information in the Spanish language, regarding two different domains, namely telecommunications services (TelecomServ dataset) and COVID-19 pandemic (COVID-19), were created. These datasets were manually constructed recovering information (news and opinions) from Cubadebate (www.cubadebate.cu), which is one of the most important and visited digital news platforms available in Cuba. For both datasets, the news-selection task was carried out while considering two fundamental requirements:

- The news should have an interest in national scope;
- The news should have more than 50 associated opinions or comments.

The TelecomServ dataset consists of 80 news and its associated opinions. Selected news are related to the Cuban Telecommunication Enterprise S.A. (ETECSA) and published in the last three years. The gathered information is one of the information sources that the enterprise may consider for measuring the customer's satisfaction regarding its services. On the other hand, the COVID-19 dataset consists of 85 news, along with their associated opinions, related to the battle against the one SARS-CoV2 coronavirus pandemic in Cuba. This dataset mostly gathers news related to information emitted by government authorities that were published in six months of the pandemic (March–August 2020). In this case, the gathered information and its processing/summarizing could be of great value for monitoring the social impact of the government actions for breaking the pandemic growth and the events that emerge in this difficult situation. The characterization of these datasets is shown in Table 1.

Table 1. Dataset characterization.

Datasets/Characteristics	#News	#Opinions	#Opinions/News	#Sentences	#Sentences/Opinion	#Terms	#Terms/Opinion
TelecomServ	80	15,776	197.2	34,665	2.2	917,674	58.2
COVID-19	85	21,707	255.4	55,447	2.5	1,587,813	73.1

4.2. Evaluation Metrics

Evaluation in text summarization can be extrinsic or intrinsic. In an extrinsic evaluation, summaries are assessed in the context of a specific task a human or machine has to carry out. In an intrinsic evaluation, summaries are evaluated about some ideal model. An intrinsic evaluation has been the most adopted paradigm, and ROUGE (Recall-Oriented Understudy for Gisting Evaluation) measures [43] are the most widely used metrics for evaluating automatic summaries. However, these content-based evaluation metrics require us to compare the automatic summary with a human summary model; this is a problem when this human summary is not available.

The effectiveness of our proposal was evaluated in a real context where the human summary model is not available; therefore, the ROUGE measures would be discarded. To address this problem, we use Jensen–Shannon divergence [16] as the quality evaluation metric for assessing our automatic summary from different perspectives. The adoption of this metric is mainly motivated by two reasons: (1) good summaries to be characterized by a low divergence between probability distributions of words in the input and summary would be expected [44] and (2) several reported studies demonstrate the existence of a strong correlation among measures that use human models (e.g., ROUGE, Pyramids, and others) and the Jensen–Shannon metric [44,45]. These studies and their experiments were developed in the context of generic multi-document summarization, topic-based multi-document summarization [44], and opinion summarization tasks [45].

Jensen–Shannon divergence (*JSD*) is an Information-Theoretic measure of divergence between two probability distributions and is defined as shown in Equations (8)–(10) [45]:

$$JSD(P \parallel D) = \frac{1}{2} \sum_{w} P_w \log_2 \frac{2P_w}{P_w + Q_w} + Q_w \log_2 \frac{2Q_w}{P_w + Q_w} \tag{8}$$

$$P_w = \frac{C_w^T}{N} \tag{9}$$

$$Q_w = \begin{cases} \frac{C_w^S}{N_S} & if \ w \in S \\ \frac{C_w^T + \delta}{N + \delta * B} & otherwise \end{cases} \tag{10}$$

where P is the probability distribution of a word, w, in the text, T, and Q is the probability distribution of a word, w, in a summary, S; N, defined as $N = N_T + N_S$, is the number of words in the text (N_T) and the summary (N_S); B is equal to $1.5\,|V|$, where V is the vocabulary extracted from the text and the summary; C_w^T is the number of words, w, in the text; and C_w^S is the number of words, w, in the summary. For smoothing the summary's probabilities, we used $\delta = 0.005$. The *JSD* measure values are in the range [0, 1], where a lower value indicates a low divergence between the compared two probability distributions, resulting in a better quality of the automatic summary in our context. This measure can be applied to the distribution of units in system summaries P and reference summaries Q, and the value obtained would be used as a score for the system summary [45]. Nevertheless, in our evaluation framework, this measure was applied according to Reference [44], using the input (text news and opinions set) as a reference, through comparing the distribution of words in full input documents with the distribution of words in automatic summaries.

Topic detection constitutes another key piece in our summarization framework; therefore, its evaluation is also very important. The proposed topic-detection process was conceived through a clustering approach, applying a HAC algorithm, which suggests that, the higher quality the clustering process has, the higher quality the topic detection has. According to this supposition, we decide to apply the Silhouette measure [15]. Silhouette, a clustering validity measure, is conceived to select the optimal number of clusters with ratio scale data (as in the case of Euclidean distances) that are suitable for a separated cluster. It is important to point out that Silhouette values range from −1 to +1, where a high value indicates that the object is well matched to its cluster and poorly matched to neighboring clusters, therefore resulting in a better quality of the clustering process.

4.3. Experimental Setup

In this section, we describe the experimental setup that was considered for both datasets and used to evaluate the effectiveness of the proposed news focused opinion summarization model. In our experiments, several solutions based on our model were developed and evaluated, to identify the best alternatives. The characterization of the evaluated approaches is shown in Table 2. For each piece of processed news and automatically generated summary with each of these solutions, we computed the averaged Silhouette and *JSD* measures. The *JSD* measure was computed from two perspectives:

- To measure the divergence between the automatic summary and the news content (*JSD focused on the news*), intending to know the correspondence level of the generated summary concerning the news.

- To measure the divergence between the automatic summary and the content of all opinions (*JSD focused on opinions*), intending to know the correspondence level of the generated summary concerning all opinions. The generated summary not only should be relevant to the news, but it should also be a good synthesis of the opinion set.

Table 2. Characterization and identification of the evaluated solutions.

| Topic Detection Approaches | Semantic Processing Based on WordNet | | | Semantic Processing Based on Word Embeddings | | |
| | Relevance Scoring | | | Relevance Scoring | | |
	Explanatoriness Scoring	TextRank Scoring (Baseline)	Sentence-to -News Scoring	Explanatoriness Scoring	TextRank Scoring (Baseline)	Sentence-to -News Scoring
Term clustering	OS1-WN	OS3-WN	OS4-WN	OS1-we	OS3-we	OS4-we
Sentence Clustering	OS2-WN	OS5-WN	OS6-WN	OS2-we	OS5-we	OS6-we

The following experimental tasks were performed:

1. Evaluating two topic detection approaches by using both term and sentence based granularities in the clustering process and comparing them by applying both WordNet

and word-embedding-based semantic-processing approaches. Selecting the clustering and semantic-processing approaches that provide the best results for topic detection.

2. Evaluating the automatically generated summaries from each solution in Table 2 according to *JSD focused on the news* (*JSD*$_{News}$) and *JSD focused on opinions* (*JSD*$_{Opinions}$), considering both WordNet and word-embeddings-based semantic-processing approaches. The obtained results would provide more details to the evaluation of the different configurations of the proposed model.

3. Comparing the results obtained by each solution in the previous tasks, identifying the best alternative for news-focused opinion summarization. TextRank-based [41] solutions are adopted as a baseline to evaluate the generated summaries according to the *JSD* measure. The best solution based on our model should work better than this popular and standard text summarization method.

Wilcoxon's Statistics Test was performed to validate the obtained results and to find significant differences between the evaluated solutions. From each dataset, 100% news and opinions were selected to constitute the sample group. In each test, the statistical significance was 95%, which means that the null hypothesis (H_0) will be rejected when the p-value ≤ 0.05.

4.4. Results and Discussion

Figures 2 and 3 show detailed results of the first experimental task, where the evaluated solutions are grouped by the clustering approaches (term and sentence clustering), and the semantic processing (WordNet or word embeddings). This experimental task is focused on the Silhouette measure. Figures 4 and 5 show a comparative summary of the averaged Silhouette values for both datasets.

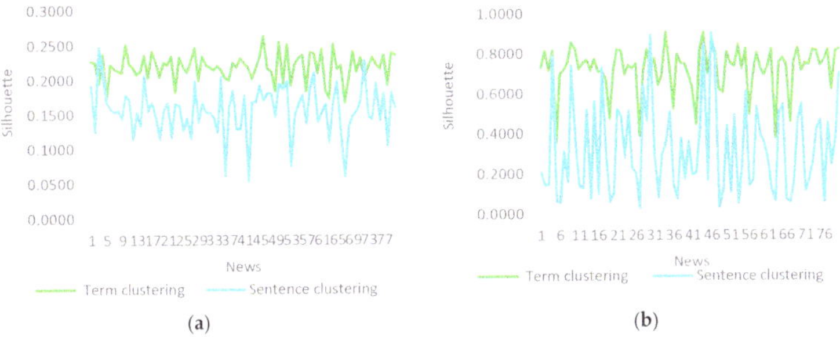

Figure 2. Results of the Silhouette measure for the two clustering approaches in the topic detection on the TelecomServ dataset by applying (**a**) WordNet and (**b**) word embeddings based semantic processing approaches.

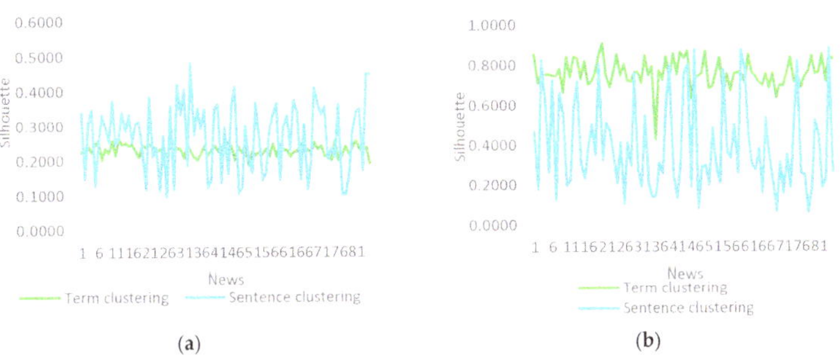

Figure 3. Results of the Silhouette measure for the two clustering approaches in the topic detection on COVID-19 dataset by applying (**a**) WordNet and (**b**) word embeddings based semantic processing approaches.

As shown in Figures 2 and 3, Silhouette values are generally better when terms are clustered, regardless of the used semantic processing technique. Only in the case of the COVID-19 dataset, when WordNet is used (Figure 3a), do Silhouette values show better performance when sentences are clustered. It is important to point out that Silhouette values associated with each news show less dispersion when term clustering is applied, which is very positive behavior, because that means it is less sensitive to the diversity of news length and the number of associated opinions. Besides, term clustering represents a more stable clustering quality behavior. According to Figures 4 and 5, applying word embedding representation reaches best-averaged Silhouette values, those that are significantly higher when terms are clustered. These results allow us to conclude that term clustering, combined with word embeddings, is a more promising and effective setting of the topic modeling in our model. This combination guarantees good quality in the clustering-based topic detection, under the assumption that the quality of the detected topics is proportional to the clustering quality.

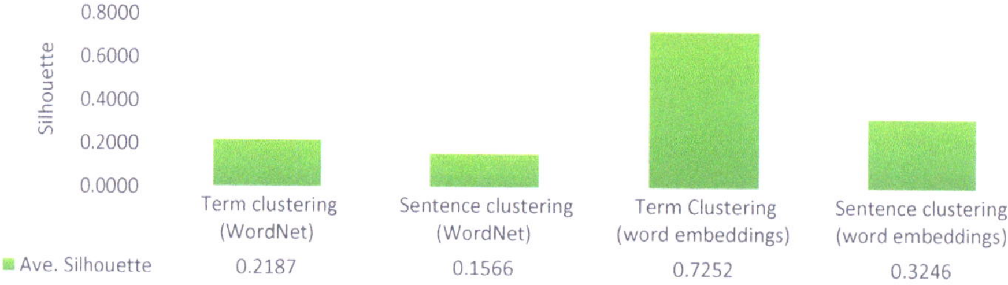

Figure 4. Averaged Silhouette values of compared topic detection approaches applied to the TelecomServ dataset.

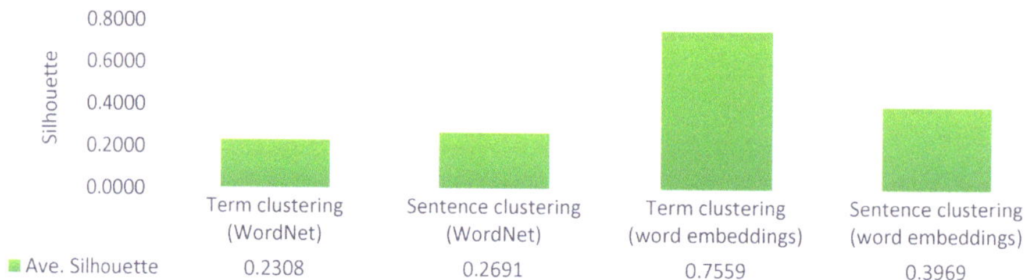

Figure 5. Average Silhouette values of compared topic detection approaches applied to the COVID-19 dataset.

Figures 6–9 show the detailed results associated with the second experimental task, which is based on the *JSD* measure. The evaluated and compared solutions are grouped according to the *JSD* scope focused on news or all opinions, as well as both term and sentence clustering. The semantic processing approach is specified in the identification of each solution (according to Table 2), which allows for an integral analysis of all developed model instances. As shown in Figures 6–9, OS4-WN and OS4-we are solutions that obtained the best results from JSD_{News} in both datasets, concerning the use of WordNet (OS4-WN) or word embeddings (OS4-we). These results indicate that combining topic modeling based on term clustering with the proposed *Sentence-to-news_scoring* for the sentence ranking is the setting of our model that allows us to generate automatic summaries more aligned to the main topics in the news, regardless of the semantic processing approach adopted.

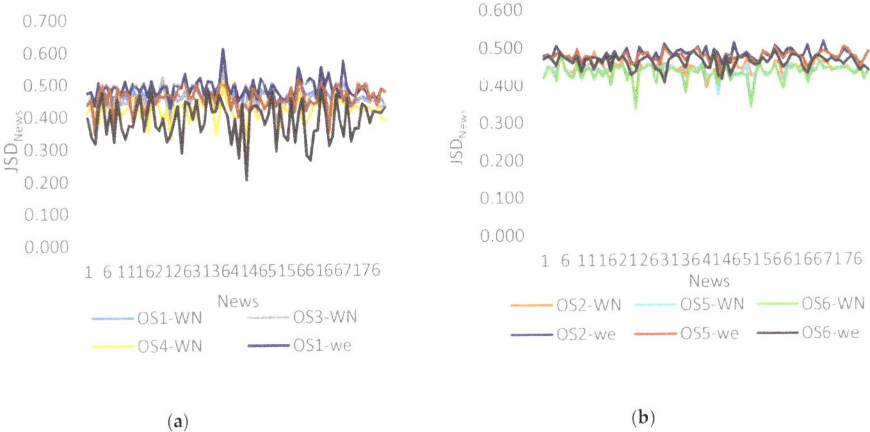

(**a**)　　　　　　　　　　　　(**b**)

Figure 6. Results of *JSD$_{News}$* (Jensen–Shannon divergence focused on the news) applying (**a**) term and (**b**) sentence clustering, using WordNet and word embeddings on the TelecomServ dataset.

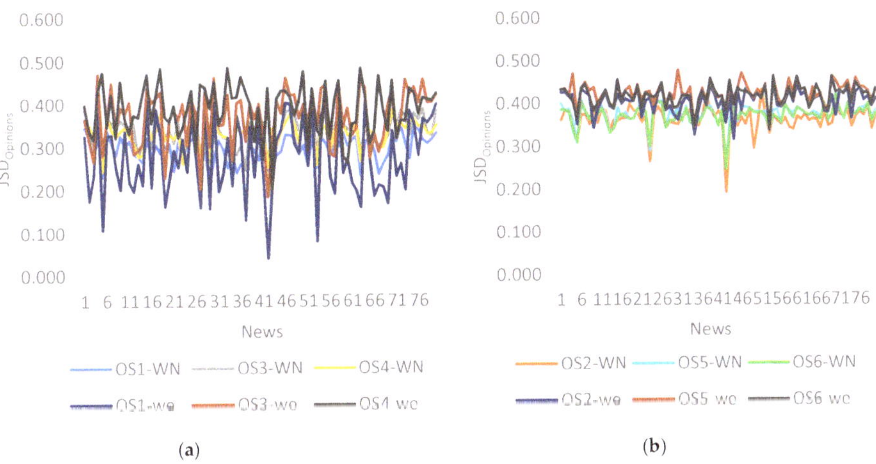

(**a**)　　　　　　　　　　　　(**b**)

Figure 7. Results of *JSD$_{Opinions}$* (Jensen–Shannon divergence focused on the opinions) applying (**a**) term and (**b**) sentence clustering, using WordNet and word embeddings on the TelecomServ dataset.

On the other hand, OS1-WN and OS1-WN are solutions that reach the best results from *JSD$_{Opinions}$* in both datasets, which means that *Explanatoriness_scoring* reaches better effectiveness to summarize the most important ideas of all opinions. These solutions do not ensure that the generated summaries have higher alignment with the news, concerning other solutions. Nevertheless, *JSD* focused on news obtained by these solutions, and their comparison with the rest of the solutions (see Tables 3 and 4) suggests that the inclusion of the topic-contextualization phase in the proposed model improves news-focused opinion summarization. Unlike the results shown in the first experiment, sentence clustering shows less sensitive behavior concerning the diversity of news length and the number of associated opinions.

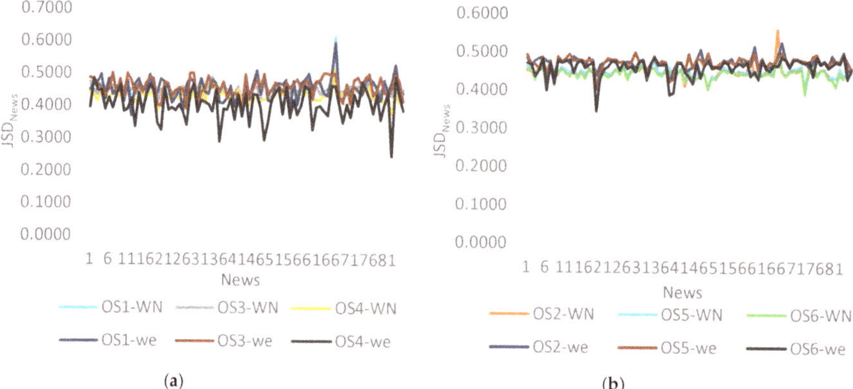

Figure 8. Results of JSD_{News} applying (**a**) term and (**b**) sentence clustering, using WordNet and word embeddings on the COVID-19 dataset.

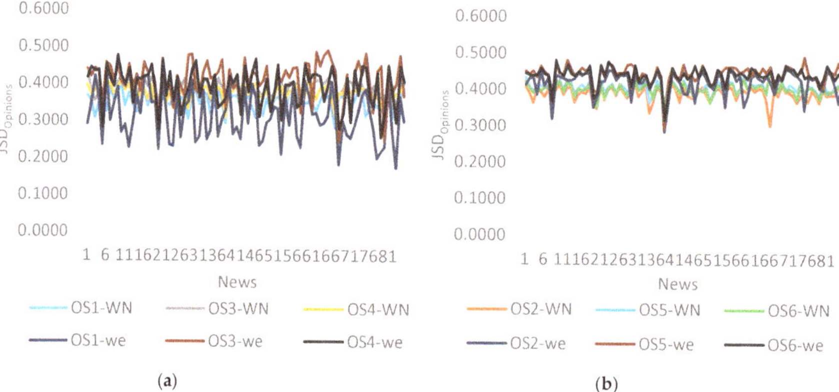

Figure 9. Results of $JSD_{Opinions}$ applying (**a**) term and (**b**) sentence clustering, using WordNet and word embeddings on the COVID-19 dataset.

Results shown in Tables 3 and 4, as well as in Figures 6–9, signify that the combination of term clustering and the word embedding representation model is also the more promising and effective setting of our model for reaching news-focused automatic summaries. Tables 3 and 4 show the averaged results of the JSD_{News} and $JSD_{Opinions}$ metrics, allowing them to complete the objective of the third task. Results of the WordNet-based semantic processing approaches are shown in Table 3, where OS3-WN was adopted as baseline 1. Results of the word-embedding-based semantic processing approaches are shown in Table 4, where OS3-we was adopted as baseline 2. These baselines were selected because the previous evaluation task concludes that the term clustering is the more promising and effective setting for topic modeling in our proposal. Thus, it allows us to evaluate the performance of the different approaches of our model and to compare them with notable summarizers as TextRank [41] (a similar decision is adopted in References [46,47]).

All solutions are compared according to the JSD scope for both datasets, and the best results are highlighted in bold. This comparison allows us to have a better understanding of the behavior of each approach. In general, the obtained results also showed that OS4-we is the best setting of our proposed model, according to JSD_{News} in both datasets. Furthermore, OS4-we is one of those solutions with best results from $JSD_{Opinions}$ when the word embedding representation is applied. This result allows us to conclude that the integration of term clustering, word embeddings, and the similarity-based sentence-to-news scoring turned out to be the more promising and effective setting of our model. The automatic summaries obtained with OS4-we are more focused on the news content; they also cover the main topics in the opinion set, reaching an appropriate balance among these targets.

Table 3. Summary of averaged results of the JSD_{News} and $JSD_{Opinions}$ metrics considering WordNet-based semantic processing.

Compared Solutions	TelecomServ		COVID-19	
	$JSD_{Opinions}$	JSD_{News}	$JSD_{Opinions}$	JSD_{News}
OS1-WN	0.296	0.465	0.351	0.448
OS2-WN	0.361	0.443	0.390	0.447
OS4-WN	0.331	0.418	0.376	0.431
OS5-WN	0.374	0.435	0.408	0.449
OS6-WN	0.369	0.430	0.403	0.443
Baseline 1: OS3-WN TextRank [41]	0.335	0.449	0.390	0.449

Table 4. Summary of averaged results of the JSD_{News} and $JSD_{Opinions}$ metrics considering word-embedding-based semantic processing.

Compared Solutions	TelecomServ		COVID-19	
	$JSD_{Opinions}$	JSD_{News}	$JSD_{Opinions}$	JSD_{News}
OS1-we	0.278	0.487	0.314	0.453
OS2-we	0.403	0.479	0.420	0.474
OS4-we	0.388	0.388	0.392	0.404
OS5-we	0.424	0.473	0.445	0.474
OS6-we	0.416	0.459	0.439	0.460
Baseline 2: OS3-we TextRank [41]	0.370	0.457	0.411	0.458

The previous results were validated through statistical tests. Wilcoxon's test was applied to find significant differences between the OS4-we results and those obtained by the rest of the evaluated solutions, using JSD_{News} as quality metrics, as shown in Table 5. The statistical results show that there are significant differences between OS4-we and the compared solutions, since the obtained p-value is less than 0.05; thus, the null hypothesis in all compared cases is rejected. On the other hand, according to the #items-best values, OS4-we obtains best results for 87% of news (as average) in the TelecomServ dataset and the 85% of news (as average) in the COVID-19 dataset. Therefore, OS4-we is the best configuration of our proposed model for news-focused opinion summarization.

Table 5. Statistical results of Wilcoxon's test from OS4-we vs evaluated solutions.

Compared Solutions	TelecomServ (80 News)				COVID-19 (85 News)			
	Statistics Variables				Statistics Variables			
	Mean-Difference	z-Value	p-Value	#Items-Best	Mean-Difference	z-Value	p-Value	#Items-Best
OS1-WN	−0.07	−7.5094	<0.00001	76	−0.04	−6.6506	<0.00001	69
OS2-WN	−0.06	−6.052	<0.00001	67	−0.04	−6.6377	<0.00001	67
OS3-WN	−0.07	−6.4606	<0.00001	66	−0.04	−6.7884	<0.00001	72
OS4-WN	−0.05	−3.6639	<0.00043	52	−0.03	−4.8421	<0.00001	55
OS5-WN	−0.06	−5.1576	<0.00001	61	0.05	6.8702	<0.00001	72
OS6-WN	−0.05	−4.4902	<0.00001	61	−0.05	−6.375	<0.00001	67
OS1-we	−0.09	−7.9135	<0.00001	80	−0.06	−7.4688	<0.00001	80
OS2-we	−0.09	−7.809	<0.00001	79	−0.06	−7.9639	<0.00001	81
OS3-we	−0.06	−7.5053	<000001	76	−0.08	−7.9553	<0.00001	81
OS5-we	−0.09	−7.6592	<0.00001	76	−0.06	−7.9209	<0.00001	82
OS6-we	−0.08	−7.2742	<0.00001	73	−0.06	−7.3869	<0.00001	73
Average	−0.07	-	-	70	−0.05	-	-	73

4.5. Illustrative Examples

Examples 1 and 2 were selected to illustrate the summaries generated by applying OS4-we on opinions about two news articles related to COVID-19, which facilitates a better understanding of how our proposal works.

Example 1. Excerpt from the summary generated regarding opinions related to the news "VALIENTES: Cuatro heroínas en la batalla contra la COVID-19" by applying OS4-we.

	News title: VALIENTES: Cuatro heroínas en la batalla contra la COVID-19		
	URL: http://www.cubadebate.cu/noticias/2020/03/30/cuatro-heroinas-en-la-batalla-contra-la-covid-19-fotos/		
Context	News fragment: A Celeste, Claudia, Esther y Melisa solo se les puede ver a través de un cristal en el Instituto de Medicina Tropical "Pedro Kourí" (IPK Cuba) y después de someterse a un complejo protocolo de seguridad (…) Ellas comparten 24 horas seguidas con la COVID-19 y necesitan una alta concentración, pues el virus pasa por sus manos y no se pueden equivocar (…) Gracias a ese arriesgado trabajo, cada día se sabe si una persona en Cuba padece o no de una pandemia que amenaza a toda la humanidad. Lo mismo ocurre en otros dos laboratorios en Villa Clara y Santiago de Cuba.		
Terms topic	'agradecerles', 'salud', 'héroe'		
Opinions	Total: 171; Sentences: 347	Pos. Score	Neg. Score
Summary	Felicitaciones a todos los que están trabajando en la epidemia del coronavirus.	1.25	1.0
	Gracias, respeto, admiración, se merecen todo nuestros médicos, todo el personal de la salud y fuera de ella que esta dando todo para erredicar este virus.	7.1	1.9
	Combatientes por la humanidad¡.	1.9	1.4
JSD$_{Opinions}$	0.374		
JSD$_{News}$	0.382		

Example 2. Excerpt from the summary generated regarding opinions related to the news "Cuba frente a la COVID-19, día 100: Últimas noticias" by applying OS4-we.

	News title: Cuba frente a la COVID-19, día 100: Últimas noticias		
	URL: http://www.cubadebate.cu/noticias/2020/06/18/cuba-frente-a-la-covid-19-dia-100-ultimas-noticias/		
Context	News fragment: Cuba entra hoy, excepto La Habana y Matanzas, en la primera fase de la recuperación de la COVID-19. El presidente Miguel Díaz-Canel subrayó este miércoles la necesidad de intensificar en ambas provincias el trabajo para que, en el menor tiempo posible, también puedan pasar a la etapa pospandemia (…) Cuando se ha dispuesto el tránsito a la primera fase de la primera etapa pos-COVID-19, en 13 provincias de la Isla y el Municipio Especial Isla de la Juventud, Matanzas y La Habana figuran como las dos únicas dolorosas excepciones que por ahora no podrán retornar a la normalidad (…) Eliminar o mantener las restricciones (tránsito paulatino de una etapa a otras) responde a criterios sanitarios y no políticos, ha explicado Torres Iríbar (…) La tasa de incidencia acumulada es de 57,5 por 100 000 habitantes, con siete municipios por encima de la media provincial: Cotorro, Centro Habana, Cerro, Regla, La Habana del Este, La Lisa y La Habana Vieja (…)		
Terms topic	'habanero', 'provincia', 'fase', 'etapa', 'indisciplina'		
Opinions	Total: 70; Sentences: 225	Pos. Score	Neg. Score
Summary	Como habanero, me siento muy apenado de que el epicentro actual y cola de la epidemia de covid 19 en cuba sea debido al comportamiento de los pobladores en mi provincia.	4.5	12.4
	Soy habanero y siento lo que diré, lo que es una pena, pero con el anuncio de que matanzas y la habana son las únicas provincias que no entran en la fase 1 de la etapa recuperativa parece que esperan compulsar a los pobladores de la habana a disciplinarse para poder llegar a esa etapa cunado la tendencia de los últimos tiempos es exactamente lo contrario de cada vez mas indisciplina.	12.1	14.5
	Veo como va en aumento las personas en las calles y la indisciplina en general como no uso o el mal uso del nasobuco, las aglomeraciones, las personas en las calles	14.7	15.6
JSD$_{Opinions}$	0.255		
JSD$_{News}$	0.357		

In these examples, some fragments of the news and generated summaries were included to avoid further extension. These examples show summaries constituted by negatives and positives sentences, as well as the terms related to the most relevant opinion topics. Terms that more contribute to compute the polarity ratings (according to the SpanishSentiWordNet lexicon) are highlighted. Selected examples illustrate that the generated summaries are strongly related to the general meaning of the news content, still when the terminology used in both information units is different. The semantic relatedness with the most relevant identified topics is also appreciated. These results are achieved due to the semantic processing conceived in our model, which is carried out by integrating a semantic representation model (word2vec [14]) and two semantic similarity measures (Wu and Palmer [36] and the sentence-to-sentence similarity measure reported in Reference [37]).

Some sentences in the generated summaries are slightly extensive, which is fundamentally due to the opinion size is not restricted in the news platform used as opinion source—being another challenge to determine the relevance of the sentences with effectiveness. The longest sentences have more probability of obtaining higher relevance scores, since they can contain a higher number of terms semantically related to the news' content. Therefore, this suggests considering other sentence features, such as tf-idf and sentence length, and their integration to the sentence relevance assessment [48].

5. Conclusions and Future Works

In this paper, we have presented a news-focused opinion summarization approach that was designed according to the conception of extractive and topic-based text summarization methods. The proposed model can retrieve relevant sentences for the essential aspects of the news (context of interest), as well as cover the main topics of the opinionated texts in the generated summary. Our proposal integrates topic modeling, sentiment analysis, news-focused relevance scoring, and semantic analysis techniques. Several techniques and settings of our model were developed and evaluated with Spanish news and opinions regarding two different domains. The selected texts come from a real digital news platform.

The proposed model outperforms both adopted baselines, which are based on the classical text summarization method TextRank, obtaining automatic summaries more relevant to the news content, as well as covering the main topics in the opinionated texts well. The integration of term clustering, word embeddings, and similarity-based-sentence-to-news scoring turned out to be the more promising and effective setting of our model, due to its reaching the best values of Jensen–Shannon divergence concerning the news and very good values for all opinions. The use of semantic representation of words for applying similarity metrics was especially effective, resulting in the best option when the word embedding representation is used. Filtering the topics non-related with the news was a crucial step for generating automatic summaries aligned with the news, as well as the calculation of the semantic similarities of the sentences with the news to extract relevant sentences. The application of the explanatoriness-scoring technique in the sentences-ranking phase reached summaries that best cover the main topics in the opinionated texts. Nevertheless, it is necessary to point out that an important factor to achieve those good results was the integration of the topic-contextualization process, where the news is used to refine the identified topics from opinions. These results give us an idea that generally the topics treated in opinions are, in fact, closely related to a context that originates them (e.g., the news).

Despite promising results, several tasks could be considered as future works. Studying the effects of applying other clustering algorithms and similarity measures could contribute to obtaining better results. In the case that there are too-short sentences, to explore opinion and sentence augmentation could improve the opinion summarization process. Besides, it would be necessary to address the problem of the inverse polarity caused by the negation and integrate several sentiment lexicons in the sentiment analysis process. The use of other sentence features and the aggregation of their results for improving the relevance scoring should also be studied.

Author Contributions: Conceptualization, A.S.-C., A.R.-H. and M.M.G.L.; methodology, A.S.-C. and M.M.G.L.; software, A.R.-H.; validation, A.R.-H., A.S.-C. and M.M.G.L.; formal analysis, A.S.-C., A.R.-H. and M.M.G.L.; investigation, A.R.-H., A.S.-C., M.M.G.L., L.A. and J.S.-G.; resources, A.R.-H.; data curation, A.S.-C. and A.R.-H.; writing—original draft preparation, A.S.-C. and A.R.-H.; writing—review and editing, A.S.-C., A.R.-H., M.M.G.L., L.A. and J.S.-G.; visualization, A.S.-C.; supervision, A.S.-C. and M.M.G.L.; project administration, A.S.-C.; funding acquisition, A.S.-C. All authors have read and agreed to the published version of the manuscript.

Funding: This research received no external funding.

Acknowledgments: The work within the projects SAFER—PID2019-104735RB-C42 (AEI/FEDER, UE) and MERINET—TIN2016-76843-C4-2-R (AEI/FEDER, UE) supported by the Spanish Government.

Conflicts of Interest: The authors declare no conflict of interest.

References

1. Balahur, A.; Kabadjov, M.; Steinberger, J.; Steinberger, R.; Montoyo, A. Challenges and solutions in the opinion summarization of user-generated content. *J. Intell. Inf. Syst.* **2012**, *39*, 375–398. [CrossRef]

2. Kumar, A.; Sharma, A. Systematic Literature Review on Opinion Mining of Big Data for Government Intelligence. *Webology* **2017**, *14*, 6–47.

3. Zhao, J.; Liu, K.; Xu, L. Sentiment Analysis: Mining Opinions, Sentiments, and Emotions. *Comput. Linguist.* **2016**, *42*, 595–598. [CrossRef]

4. Sun, S.; Luo, C.; Chen, J. A review of natural language processing techniques for opinion mining systems. *Inf. Fusion* **2017**, *36*, 10–25. [CrossRef]

5. Ravi, K.; Ravi, V. A survey on opinion mining and sentiment analysis: Tasks, approaches and applications. *Knowl. Based Syst.* **2015**, *84*, 14–46. [CrossRef]

6. Moussa, M.E.; Mohamed, E.H.; Haggag, M.H. A survey on opinion summarization techniques for social media. *Futur. Comput. Inform. J.* **2018**, *3*, 82–109. [CrossRef]

7. Condori, R.E.L.; Pardo, T.A.S. Opinion summarization methods: Comparing and extending extractive and abstractive approaches. *Expert Syst. Appl.* **2017**, *78*, 124–134. [CrossRef]

8. Li, P.; Huang, L.; Ren, G.-J. Topic Detection and Summarization of User Reviews. *arXiv* **2020**, arXiv:2006.00148.

9. Rossetti, M.; Stella, F.; Zanker, M. Analyzing user reviews in tourism with topic models. *Inf. Technol. Tour.* **2015**, *16*, 5–21. [CrossRef]

10. Chakraborty, R.; Bhavsar, M.; Dandapat, S.K.; Chandra, J. Tweet Summarization of News Articles: An Objective Ordering-Based Perspective. *IEEE Trans. Comput. Soc. Syst.* **2019**, *6*, 761–777. [CrossRef]

11. Kilgarriff, A.; Fellbaum, C. WordNet: An Electronic Lexical Database. *Language* **2000**, *76*, 706. [CrossRef]

12. Kamath, U.; Liu, J.; Whitaker, J. *Deep Learning for NLP and Speech Recognition*; Springer Nature Switzerland: Cham, Switzerland, 2019.

13. Yang, H.; Luo, L.; Chueng, L.P.; Ling, D.; Chin, F. Deep Learning and Its Applications to Natural Language Processing. In *Deep Learning: Fundamentals, Theory and Applications*; Huang, K., Hussain, A., Wang, Q.-F., Zhang, R., Eds.; Springer Nature Switzerland: Cham, Switzerland, 2019; pp. 89–109.

14. Mikolov, T.; Chen, K.; Corrado, G.; Dean, J. Efficient estimation of word representations in vector space. In Proceedings of the 1st International Conference on Learning Representations (ICLR 2013), Scottsdale, AZ, USA, 2–4 May 2013.

15. Rousseeuw, P.J. Silhouettes: A graphical aid to the interpretation and validation of cluster analysis. *J. Comput. Appl. Math.* **1987**, *20*, 53–65. [CrossRef]

16. Lin, J. Divergence measures based on the Shannon entropy. *IEEE Trans. Inf. Theory* **1991**, *37*, 145–151. [CrossRef]

17. Allahyari, M.; Pouriyeh, S.; Assefi, M.; Safaei, S.; Trippe, E.D.; Gutierrez, J.; Kochut, K. Text Summarization Techniques: A Brief Survey. *Int. J. Adv. Comput. Sci. Appl.* **2017**, *8*, 397–405. [CrossRef]

18. Abualigah, L.M.; Bashabsheh, M.Q.; Alabool, H.; Shehab, M. Text Summarization: A Brief Review. In *Recent Advances in NLP: The Case of Arabic Language, Studies in Computational Intelligence*; Abd El Aziz, M., Al-qaness, M.A.A., Ewees, A.A., Dahou, A., Eds.; Springer: Cham, Switzerland, 2020; pp. 1–15.

19. Gambhir, M.; Gupta, V. Recent automatic text summarization techniques: A survey. *Artif. Intell. Rev.* **2017**, *47*, 1–66. [CrossRef]

20. Amplayo, R.K.; Lapata, M. Informative and Controllable Opinion Summarization. *arXiv* **2019**, arXiv:1909.02322.

21. Lloret, E.; Boldrini, E.; Vodolazova, T.; Martínez-Barco, P.; Muñoz, R.; Palomar, M. A novel concept-level approach for ultra-concise opinion summarization. *Expert Syst. Appl.* **2015**, *42*, 7148–7156. [CrossRef]

22. Mukherjee, R.; Peruri, H.C.; Vishnu, U.; Goyal, P.; Bhattacharya, S.; Ganguly, N. Read what you need: Controllable Aspect-based Opinion Summarization of Tourist Reviews. In Proceedings of the 43rd International ACM SIGIR Conference on Research and Development in Information Retrieval, Xi'an, China, 25–30 July 2020; pp. 1825–1828. [CrossRef]

23. Jiang, Y.; Meng, W.; Yu, C. Topic Sentiment Change Analysis. In Proceedings of the Machine Learning and Data Mining in Pattern Recognition, MLDM 2011, New York, NY, USA, 30 August–3 September 2011; LNCS 6871. Springer: Berlin/Heidelberg, Germany, 2011; pp. 443–457. [CrossRef]

24. Ali, S.M.; Noorian, Z.; Bagheri, E.; Ding, C.; Al-Obeidat, F. Topic and sentiment aware microblog summarization for twitter. *J. Intell. Inf. Syst.* **2018**, *54*, 129–156. [CrossRef]

25. Rohit, S.V.K.; Shrivastava, M. Using Argumentative Semantic Feature for Summarization. In Proceedings of the 2019 IEEE 13th International Conference on Semantic Computing (ICSC), Newport Beach, CA, USA, 30 January–1 February 2019; pp. 456–461.

26. Abdi, A.; Shamsuddin, S.M.; Aliguliyev, R.M. QMOS: Query-based multi-documents opinion-oriented summarization. *Inf. Process. Manag.* **2018**, *54*, 318–338. [CrossRef]

27. Wang, L.; Raghavan, H.; Cardie, C.; Castelli, V. Query-Focused Opinion Summarization for User-Generated Content. In Proceedings of the COLING 2014, the 25th International Conference on Computational Linguistics, Dublin, Ireland, 23–29 August 2014; Dublin City University and Association for Computational Linguistics. pp. 1660–1669.

28. Conrad, J.G.; Leidner, J.L.; Schilder, F.; Kondadadi, R. Query-based opinion summarization for legal blog entries. In Proceedings of the 12th International Conference on Extending Database Technology Advances in Database Technology-EDBT '09, New York, NY, USA, 8–12 June 2009; pp. 167–176.

29. Luo, W.; Zhuang, F.; He, Q.; Shi, Z. Exploiting relevance, coverage, and novelty for query-focused multi-document summarization. *Knowl. Based Syst.* **2013**, *46*, 33–42. [CrossRef]

30. Ramón Hernández, A.; García Lorenzo, M.M.; Simón-Cuevas, A.; Arco, L.; Serrano-Guerrero, J. A semantic polarity detection approach: A case study applied to a Spanish corpus. *Procedia Comput. Sci.* **2019**, *162*, 849–856. [CrossRef]

31. Pennington, J.; Socher, R.; Manning, C. Glove: Global Vectors for Word Representation. In Proceedings of the 2014 Conference on Empirical Methods in Natural Language Processing (EMNLP); Association for Computational Linguistics (ACL), Doha, Qatar,, 25–29 October 2014; Association for Computational Linguistics: Stroudsburg, PA, USA, 2014; pp. 1532–1543. [CrossRef]

32. Verberne, S.; Krahmer, E.; Wubben, S.; Bosch, A.V.D. Query-based summarization of discussion threads. *Nat. Lang. Eng.* **2019**, *26*, 3–29. [CrossRef]

33. Angioni, M.; Devola, A.; Locci, M.; Tuveri, M.L.A.F. An Opinion Mining Model Based on User Preferences. In Proceedings of the 18th International Conference on WWW (Internet 2019), IADIS-International Association for the Development of the Information Society, Cagliari, Italy, 7–8 November 2019; pp. 183–185. [CrossRef]

34. Dalal, M.K.; Zaveri, M.A. Semisupervised Learning Based Opinion Summarization and Classification for Online Product Reviews. *Appl. Comput. Intell. Soft Comput.* **2013**, *2013*, 1–8. [CrossRef]

35. Manning, C.; Prabhakar, R.; Schütze, H. *An Introduction to Information Retrieval*; Cambridge University Press: Cambridge, UK, 2008.

36. Pedersen, T.; Patwardhan, S.; Michelizzi, J. WordNet:Similarity-Measuring the Relatedness of Concepts. In Proceedings of the 19th National Conference on Artificial Intelligence (AAAI-04), San Jose, CA, USA, 25–29 July 2004; pp. 1024–1025.

37. Mihalcea, R.; Corley, C.; Strapparava, C. Corpus-based and Knowledge-based Measures of Text Semantic Similarity. In Proceedings of the 21st National Conference on Artificial Intelligence (AAAI'06), Boston, MA, USA, 16–20 July 2006; pp. 775–780.

38. Baccianella, S.; Esuli, A.; Sebastiani, F. SentiWordNet 3.0: An enhanced lexical resource for sentiment analysis and opinion mining. In Proceedings of the 7th International Conference on Language Resources and Evaluation, Valletta, Malta, 17–23 May 2010; pp. 2200–2204.

39. Amores, M.; Arco, L.; Borroto, C. Unsupervised Opinion Polarity Detection based on New Lexical Resources. *Comput. Sist.* **2016**, *20*, 263–277. [CrossRef]

40. Kim, H.D.; Castellanos, M.G.; Hsu, M.; Zhai, C.; Dayal, U.; Ghosh, R. Ranking explanatory sentences for opinion summarization. In Proceedings of the 36th International ACM SIGIR Conference on Research and Development in Information Retrieval-SIGIR '13, Dublin, Ireland, 28 July–1 August 2013; p. 1069.

41. Mihalcea, R.; Tarau, P. TextRank: Bringing Order into Texts. In Proceedings of the 2004 Conference on Empirical Methods in Natural Language Processing (EMNLP'04), Barcelona, Spain, 25–26 July 2004; pp. 404–411.

42. Brin, S.; Page, L. The anatomy of a large-scale hypertextual Web search engine. *Comput. Networks ISDN Syst.* **1998**, *30*, 107–117. [CrossRef]

43. Lin, C.-Y. Rouge: A Package for Automatic Evaluation of Summaries. In Proceedings of the Text Summarization Branches Out, Barcelona, Spain, 25–26 July 2004; pp. 74–81.

44. Louis, A.; Nenkova, A. Automatically evaluating content selection in summarization without human models. In Proceedings of the 2009 Conference on Empirical Methods in Natural Language Processing, Singapore, 6–7 August 2009; pp. 306–314.

45. Saggion, H.; Torres-Moreno, J.M.; da Cunha, I.; SanJuan, E. Multilingual Summarization Evaluation without Human Models. In *Proceedings of the Coling 2010: Poster*; Beijing, China, 23–27 August 2010, Association for Computational Linguistics: Stroudsburg, PA, USA, 2010; pp. 1059–1067.

46. Coavoux, M.; Elsahar, H.; Gallé, M. Unsupervised Aspect-Based Multi-Document Abstractive Summarization. In Proceedings of the 2nd Workshop on New Frontiers in Summarization, Hong Kong, China, 3–4 November 2019; pp. 42–47.

47. Elsahar, H.; Coavoux, M.; Gallé, M.; Rozen, J. Self-Supervised and Controlled Multi-Document Opinion Summarization Hady. *arXiv* **2020**, arXiv:2004.14754.

48. Valladares-Valdés, E.; Simón-Cuevas, A.; Olivas, J.A.; Romero, F.P. A Fuzzy Approach for Sentences Relevance Assessment in Multi-document Summarization. In *International Workshop on Soft Computing Models in Industrial and Environmental Applications*; Springer: Cham, Switzerland, 2019; pp. 57–67.

Publisher's Note: MDPI stays neutral with regard to jurisdictional claims in published maps and institutional affiliations.

 *information*

Article

Sentiment Analysis and Text Mining of Questionnaires to Support Telemonitoring Programs

Chiara Zucco [1], Clarissa Paglia [2], Sonia Graziano [3], Sergio Bella [2] and Mario Cannataro [1,4,*

[1] Department of Medical and Surgical Sciences, University Magna Graecia of Catanzaro, 88100 Catanzaro, Italy; chiara.zucco@unicz.it

[2] Unit of Cystic Fibrosis, Bambino Gesù Children's Hospital, 00165 Rome, Italy; clarissa.paglia@opbg.net (C.P.); sergio.bella@opbg.net (S.B.)

[3] Unit of Clinical Psychology, Bambino Gesù Children's Hospital, 00165 Rome, Italy; graziano.sonia@gmail.com

[4] Data Analytics Research Center, University Magna Graecia of Catanzaro, 88100 Catanzaro, Italy

* Correspondence: cannataro@unicz.it

Received: 15 October 2020; Accepted: 24 November 2020; Published: 26 November 2020

Abstract: While several studies have shown how telemedicine and, in particular, home telemonitoring programs lead to an improvement in the patient's quality of life, a reduction in hospitalizations, and lower healthcare costs, different variables may affect telemonitoring effectiveness and purposes. In the present paper, an integrated software system, based on Sentiment Analysis and Text Mining, to deliver, collect, and analyze questionnaire responses in telemonitoring programs is presented. The system was designed to be a complement to home telemonitoring programs with the objective of investigating the paired relationship between opinions and the adherence scores of patients and their changes through time. The novel contributions of the system are: (i) the design and software prototype for the management of online questionnaires over time; and (ii) an analysis pipeline that leverages a sentiment polarity score by using it as a numerical feature for the integration and the evaluation of open-ended questions in clinical questionnaires. The software pipeline was initially validated with a case-study application to discuss the plausibility of the existence of a directed relationship between a score representing the opinion polarity of patients about telemedicine, and their adherence score, which measures how well patients follow the telehomecare program. In this case-study, 169 online surveys sent by 38 patients enrolled in a home telemonitoring program provided by the Cystic Fibrosis Unit at the "Bambino Gesù" Children's Hospital in Rome, Italy, were collected and analyzed. The experimental results show that, under a Granger-causality perspective, a predictive relationship may exist between the considered variables. If supported, these preliminary results may have many possible implications of practical relevance, for instance the early detection of poor adherence in patients to enable the application of personalized and targeted actions.

Keywords: text mining; sentiment analysis; Web-based questionnaire; telemedicine; telemonitoring; telehomecare

---

1. Introduction

Telemedicine can be defined as the set of health services providing medical care in patients' daily living environment, which is possible thanks to the support of information and telecommunication technologies [1].

Common goals of telemedicine programs are substantially threefold [2–4]:

- to increase self-management skills for patients whether they have a chronic condition or, for instance, during recovery or a rehabilitation phase after surgery or in the follow-up after a long time hospitalization and also during the treatment for depression and other mental health conditions;

- to improve patient surveillance and medication adherence by remote monitoring of their health status; and
- to reduce healthcare costs by optimizing the doctor's work by reducing the number of accesses to the emergency room and reducing the average of hospital admissions per patient.

A subfield of telemedicine is telehomecare, or home telemonitoring, which enables the rapid exchange of information between health systems and patients. Patients enrolled in a telehomecare program are provided with bio-monitoring devices and Internet reporting systems installed in their daily living environment. The devices can be used to autonomously measure vital signals, specifically related to the specific condition of the patient. These measurements are then transmitted and evaluated by health professionals (physicians and nurses) who will subsequently re-contact patients via phone call or message to check their symptoms and, eventually, provide an early medical response.

Telemedicine and, specifically, telehomecare systems have shown themselves to be cost-effective [5] and to provide an improvement in patients' quality-of-life, in terms of significant reduction of both mortality and length of stay of patients in progressive care unit [6], significant improvement of glycemic control for patients with diabetes [7], etc.

However, two variables that may affect a telemonitoring program's effectiveness are adherence levels and the degrees of drop-out. Adherence levels may be measured in different suitable ways. Here, adherence is intended as the rate of performed monitoring events with respect to the ideal number of events, suggested by the telemonitoring protocol, while the degrees of drop-out refer to the percentage of patients who abandon the telemedicine or the telehomecare program they were enrolled in, generally due to poor adherence. In particular, in [8], a systematic review among 37 healthcare programs for Heart Failure and Chronic Obstructive Pulmonary Disease was carried out. In the study, rates of refusal of almost one-third of patients were reported. Moreover, among patients who took part in the telehomecare program, one-fifth abandoned the program after enrollment.

Another interesting study is related to cystic fibrosis patients' follow-up at home [9].

Cystic fibrosis is the most common life-threatening genetic disease in the Caucasian population [10]. It is characterized by recurrent episodes of a respiratory infection that cause progressive lung deterioration, with a long-term decline in lung function. The spirometry test is a simple test used for lung function monitoring. It is known that continuous monitoring of lung function during the follow-up of patients with cystic fibrosis can reduce lung damage by preventing bronchopulmonary exacerbations and, consequently, prevent patient's exitus for lung insufficiency [11].

The authors reported that, of 39 enrolled patients, 15 dropped out of the program (38.46%). The percentage decreases to 31.4% if considering voluntary drop-out. Eighty-one percent of drop-out was due to poor program adherence [9].

The most frequently used approaches for conducting research studies in the social sciences make use of surveys [12]. Among the methods for collecting survey data, questionnaires represent a widely used tool. Thanks to the availability of tools and systems that facilitate the development and administration phases through online platforms, their popularity has grown significantly. Compared to face-to-face or telephone interviews and to questionnaires-on-paper, online questionnaires provide several advantages: (i) a cost reduction; (ii) the collection of a greater number of data in a shorter time; (iii) the possibility for the individual to manage the place and time to take the questionnaire; and (iv) responses are already digitized and exportable in formats which are suitable for a subsequent analysis [13].

Thus far, closed-ended questions in questionnaires have dominated the scene in the social sciences and, consequently, in the psychological and health sciences. This choice is justified by the easiness of data collection, reliability and simplicity of analysis, and the possibility of standardizing the collection to compare results between different populations [14]. On the other hand, the possibility of using open-ended questions would allow performing fine-grained analysis by offering new and interesting insights, capable of detecting slight differences, especially, for example, in the context of patient monitoring.

The investigations carried out to verify if a significant benefit may be obtained by introducing open-ended questions in questionnaires led to different results. In [15,16], it is shown that answers which received a high response rate in closed-ended questionnaires were not mentioned when the same question was formulated in an open-ended form, whereas the study conducted in [17] showed no benefit in introducing open-ended questions.

With the availability of many textual data coming from social platforms, noteworthy developments concerning the automated analysis of texts have been registered during the last decade. Above all, an increasing interest in the field of sentiment analysis, which aims at the automatic extraction of emotions and opinions, mainly from text [18], has been witnessed.

This work aims to present an integrated software architecture for the online provision and collection of questionnaires or surveys, which exploits a sentiment analysis-based approach to monitor patients' adherence to telehomecare programs. The idea is that the sentiment, i.e., the degrees of positiveness/negativeness, expressed by patients through their responses to questionnaires, may be related to their adherence and used to predict drop-out.

The present architecture proposal is intended as a contribution that can help the context of home telemonitoring programs. The basic idea is to integrate within a telehomecare system an online survey instrument to investigate the polarities of patients' opinions in relation to their experience.

The proposed system also encompasses a novel analysis approach that leverages lexicon-based sentiment analysis techniques and exploits the inferred polarity as a numerical feature to enhance further statistical or machine learning analysis.

To the best of our knowledge, no specific research has been published, nor has a system architecture been proposed that would explicitly monitor changes in patient's opinion across time through the repeated administration of a questionnaire, using the polarity associated with answers to open-ended question as a numerical feature, in a telehomecare system.

Additionally, the paper presents a case study application of the system architecture to discuss whether a predictive relationship, in terms of Granger-causality test modeling, may be assessed between patient adherence in a cystic fibrosis telehomecare program and their opinion about the program they are enrolled in.

The rest of the paper is organized as follows. Section 2 describes the methodology behind the proposed approach and the case-study application. Section 3 provides insights related to collected data and presents the Granger-causality hypothesis tests results and discusses it. Finally, Section 4 concludes the paper and outlines future works.

2. Materials and Methods

In this section, some preliminary information related to the case study, a description of the experimental protocol used and the analysis pipeline's proposal are presented.

2.1. Preliminary Information

Since 2001, a home telemonitoring program has been provided by the Cystic Fibrosis Center of the "Bambino Gesù" Pediatric Hospital in cystic fibrosis patients' follow-up. Patients are provided with Spirotel instrumentation from MIR (Medical International Research), which transmits data from the spirometry test and overnight pulse oximetry remotely, following the clinical workflow detailed in [11]. Patients are suggested to send spirometry transmission at least twice a week.

After the transmission, physicians contact patients by performing a telephone interview involving questions about some pulmonary symptoms and more general health conditions. Patients included in the telemonitoring program are treated with standard follow-up protocols, similar to those not enrolled in the program. A detailed description can be found in [19]. Despite the promising results, a significant percentage of abandonment for poor adherence has been constantly registered.

Table 1 reports some statistics related to patient enrollment, related to a nine-year period (2010–2018). As shown in Table 1, the drop-out patients enrolled in the telehomecare program represent 41% of the total number. Table 2 further illustrates the composition of patients who leave the program. In particular, 81.25% of patients' drop-out is due to voluntary drop-out: 50% of patient abandonment is related to poor adherence, while 31.25% of intentional abandonment is related to other reasons.

Table 1. Balance of enrolment during the period 2010–2018.

Patients	n.	Percentage
Enrolled	78	
Active	46	58.97
Drop-out	32	41.03

Table 2. Proportion of patients drop-out during the period 2010–2018, grouped by abandonment causes.

Drop-Out Patients	n.	% of Total	% of Dropout
poor adherence	16	20.51	50
died	6	7.69	18.75
other	10	12.82	31.25

2.2. Experimental Protocol and Dataset Description

The data analyzed in the present case study application were collected from the Cystic Fibrosis Unit, Bambino Gesù Children's Hospital, Rome, Italy. In this study, 169 online surveys sent by 38 cystic fibrosis patients (F/M = 20/18, age = 28.7 ± 9.91, age range = 14–49) recruited among patients already enrolled in a telemedicine program (years of enrollment = 5.9 ± 3.9) were collected and analyzed at five different survey epochs.

The enrollment criteria included patients more than 12 years old with cystic fibrosis who access the Cystic Fibrosis Unit in ordinary, daytime, or outpatient hospitalization. All patients who have undergone a transplant (liver/lung) were excluded from the study.

The study was formally approved by the local Medical Research Ethics Committee.

2.3. Administration of Questionnaire

From June 2019, 38 enrolled patients were asked to complete, every three months, an online questionnaire designed ad-hoc by the clinical team. In the following, each set of surveys submission is indicated as an epoch.

The Telemedicine Drop-Out (TDO) questionnaire consists of 15 blocks of closed, mixed, and open-ended items with yes/no constraints, and it was administered through a self-hosted web-based survey instrument built on top of LimeSurvey. The TDO survey was designed as an online, structured version of the interview led by the medical team within the telemedicine program, extended with a series of open-ended questions, whose objective was to infer polarity or, in perspective, to extract emotions from the relative answers [20]. The TDO questionnaire is reported in Appendix A.

To administer surveys to patients, LimeSurvey (https://www.LimeSurvey.org/) [21] is set up as a highly customizable, free, and responsive online survey tool. It also provides various API functions through the LimeSurvey RemoteControl 2 (LSRC2). The survey structure and the participants are created through the user interface provided by LimeSurvey. The collection of survey answers is automatized using the Python library Limepy that provides a Python wrapper for the LSRC2 API and the Python library Schedule to automatically update the responses. The DBMS server is MySQL.

As already stated, adherent patients need to transmit the results of the spirometry test at least twice a week. For each survey administration, i.e., survey epoch, the patient's adherence score (Adh-score) to the telemonitoring program was assessed as the total number of spirometry transmissions sent during a three-month window starting from the month before until the month subsequent to the survey administration, averaged by twice the total number of weeks following. More in details, suppose that a survey was carried at month t, then:

$$\text{Adh-score}_t = \frac{nS_{t-1} + nS_t + nS_{t+1}}{2(w_{t-1} + w_t + w_{t+1})}$$

where nS_{t-1}, nS_t, and nS_{t+1} refer to the number of spirometry transmissions sent in month $t-1$, t, and $t+1$, respectively, while w_{t-1}, w_t, and w_{t+1} refer to the number of weeks in months $t-1$, t, and $t+1$, respectively.

For instance, to calculate the Adh-score related to the first epoch submission, t = June 2019. Therefore, each patient's total number of spirometry transmissions from May 2019 to July 2019 was considered. Moreover, since the three-month window encompasses 13 weeks, the total number of spirometry transmissions was averaged by twenty-six.

By definition, patients who strictly follow medical advice have a related Adh-score ≥ 1. In the following, the percentage of Adh-score, i.e., Adh-score (%), is considered. Therefore,

$$0 \leq \text{Adh-score (\%)} = \text{Adh-score} * 100$$

and Adh-score (%) > 100 for patients with high rates of adherence. The clinical team provided the number of transmissions per month.

2.4. System Architecture

The system architecture encompasses three independent modules, connected in a cascade-fashion. In future works, the modules are supposed to be integrated using a unique user interface. Figure 1 shows the overall architecture for the system, which is organized as three logical levels:

- **Data collection:** Only completed survey data are automatically downloaded and cleaned to eliminate redundant information and duplicates. If the same respondent at the same administration epoch sent two surveys, the most recent one is considered.
- **Data integration:** Collected data are joined with adherence data according to a common schema. All the steps, typical of an ETL (Extraction, Transformation and Loading) approach, have been implemented through the Pandas library [22]. The integrated data table may be downloaded as a comma separate value (csv) file.
- **Data analysis:** Collected and integrated data encompass both structured information and free texts, resulting from patients' answers to the open-ended TDO questions. Therefore, this level implements different analysis pipelines, depending on the integrated data and analysis type.

2.4.1. Data Analysis Pipeline

The general pipeline for the analysis of textual data, i.e., answers to open-ended questions, involves:

- **Text preprocessing:** It includes standard NLP (Natural Language Processing) techniques, i.e., tokenization, stop word removal, and lemmatization. The preprocessing step was executed by using *SpaCy* (https://spacy.io/), a popular library for NLP in Python, which provides a set of preprocessing algorithms also for the Italian language.
- **Feature extraction:** To each open-ended free-text answer, a polarity score in the range $[-1, 1]$ was assigned through the VADER [23] lexicon-based method adapted to the Italian language and considered as a numerical feature.
- **Statistical hypothesis testing:** Data were sorted by respondents and survey submission date and, for each open-ended question, the sequence of assigned polarity was modeled as a time series, as the sequence of Adh-scores. Augmented Dickey–Fuller Test [24] was used to check for stationarity, while Granger-causality hypothesis test model [25] was examined to discuss the existence of directed causal interactions between the polarity score associated with free-text answers and adherence.
- **Data visualization:** To provide useful insights and summarize patient answers, different visualization techniques were used. In particular, preprocessed free-text answers are visualized through word clouds, while a graph shows the time-series of polarity scores at different submission epochs.

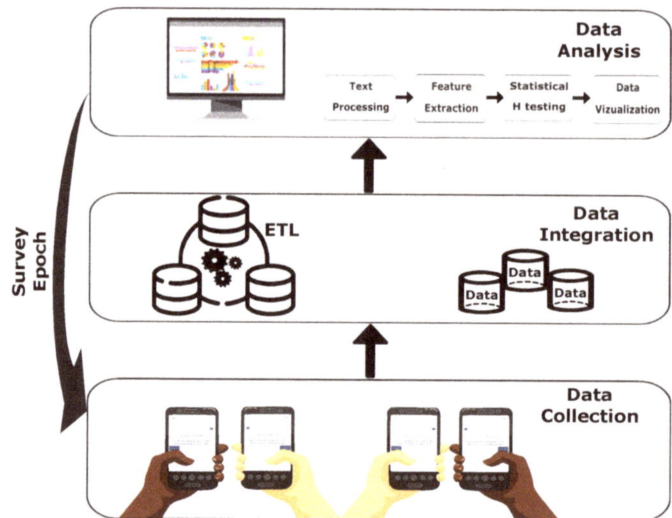

Figure 1. The modules of the system architecture, implemented as three independent levels connected to each other in cascade. The architecture is designed to be cyclical, as the system is used for each scheduled administration of the survey.

2.4.2. Sentiment Polarity Extraction

Valence Aware Dictionary for sEntiment Reasoning (VADER) [23] is a lexicon-based sentiment analysis engine that combines lexicon-based methods with a rule-based modeling consisting of five human validated rules.

The benefits of VADER's approach are: it does not require a training phase, and, consequently, its application is feasible even in low resource data domain; it works well on short text; it is fast and, therefore, may be suited for near real time application; being related on general "parsimonious" rules, it is basically domain-agnostic; and it constructs a white box model, thus is highly interpretable and adaptable to different languages.

The starting point of the VADER system is a generalizable, valence-based, human-curated gold standard sentiment lexicon, built on top of three well-established lexicons, i.e., LIWC [26], General Inquirer, and ANEW [27], expanded with a set of lexical features commonly used in social media, which include emoji, for a total of 9000 English terms subsequently annotated in a $[-4, 4]$ range through the Amazon Mechanical Turk's crowd-sourcing service.

The VADER engine's second core step is the identification of some general grammatical and syntactic heuristics to identify semantic shifters, i.e., words that increase, decrease, or change the polarity orientation of another word. In particular, five heuristics for sentiment polarity shifters have been identified:

- Punctuation: The exclamation mark (!) is a valence booster, i.e., it increases sentiment intensity without affecting sentiment orientation.
- Capitalization: Uppercase words in the presence of lower cases words should be treated as a sentiment intensifier without affecting sentiment orientation.
- Degree modifiers: Nouns, adjectives, and adverbs, as well as idioms, are known as intensifiers or down-toning, which impact sentiment intensity by increasing or decreasing it.
- Contrastive particles: The "but" conjunction between two sentences shifts sentiment polarity in favor of the second part of the text.
- Negation: Negations reverse the polarity orientation of the lexical particles they are referred to. The investigation of Trigram preceding sentiment-laden terms enables the identification of negations for that specific term.

To extend VADER to the Italian language, Sentix [28], a lexicon that automatically extends the SentiWordNet annotation to the Italian synsets provided in MultiWordNet [29], was considered.

Among the five heuristics designed in VADER, only three needed to be adapted to the Italian language since the shifter role of capitalization of words and exclamation marks is used as intensifiers for both languages. Words belonging to the VADER set of negation words were translated in the Italian language, and the set was then extended by retrieving MultiWordNet synset terms for each word, while contrastive particle "but" was simply translated to Italian.

Among the intensifier sets, VADER also considered a few idioms, but, due to discrepancies across different languages, idioms were not considered.

2.4.3. Granger-Causality Testing

Granger-causality is a statistical hypothesis testing model to determine if there is a directed relationship between two time series [25]. A time series X is said to Granger-cause Y if it can be shown that there is a statistically significant improvement in predicting future values of Y by using past values of X (i.e., lagged values of X) and Y, compared to predictions based only on past values of Y.

The possibility to relate past values of X to Y's actual values is in virtue of a lag factor. Here, the Granger-causality test was computed for X's lagged values. All the lags ranging from one to four were tested, where four is the number of considered submission epochs minus one.

Here, the considered alternative hypothesis is that the polarity-score time series associated with each considered open-ended question Granger-cause the time series of adherence. The level of significance was set at 5%, i.e., p-value < 0.05. The Granger-causality test assumes the hypothesis that the investigated time-series are stationary. Therefore, the augmented Dickey–Fuller method was exploited to check stationarity conditions [24].

3. Results and Discussion

In this section, we present the results related to the Granger-causality testing model to assess the plausibility of the existence of a predictive relationship between a score representing the opinion polarity of patients about telemedicine and their Adh-score. Moreover, to gain useful insights about the collected data, a preliminary exploratory data analysis was performed by following the pipeline discussed in the previous section and by summarizing data through suitable visualization.

3.1. Exploratory Data Analysis

In this study, 169 answers to the TDO survey were collected and analyzed following the system architecture described in the previous Section.

The present exploratory data analysis aims to provide useful insights into the data collection and integration processes.

In particular, the collected data were sent by 38 cystic fibrosis patients through five subsequent submissions, scheduled every three months on average.

Figure 2 shows a violin plot describing the distribution of Adh-score in percentage associated with each submission epoch, while, in Table 3, the same information is provided in a tabular form. Although mean values of Adh-score (%) are in the range [38.67%, 51.45%] for each epoch, the standard deviation and minimum and maximum values of Adh-score (%) show a considerable variation of Adh-values, with patients who sent zero spirometries and patients who transmitted three times more than the medical advice.

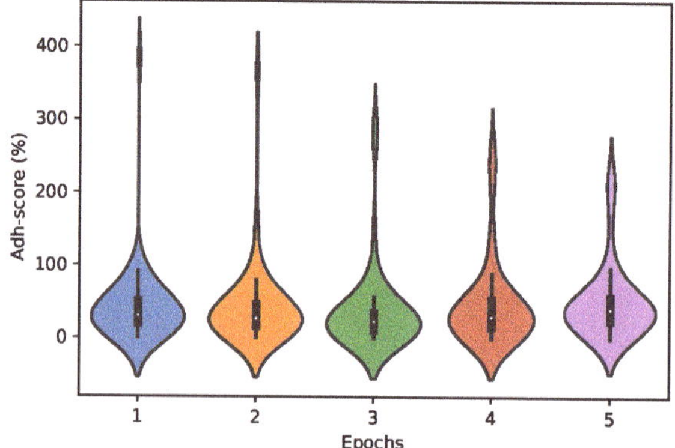

Figure 2. Distribution of Adh-score in percentage across the five subsequent submission epochs.

Table 3. Adh-score (%): descriptive statistics across five subsequent epochs.

	Epoch_1	Epoch_2	Epoch_3	Epoch_4	Epoch_5
mean	42.67	41.77	38.67	47.39	51.45
std	57.25	57.70	59.64	59.54	51.82
min	0.00	0.00	0.00	0.00	0.00
max	384.62	365.76	293.55	260.96	229.50

A comprehensive analysis of the responses to the TDO survey is beyond the scope of this paper. Instead, only answers to two open-ended questions collected from the TDO survey are discussed:

- **Q1** - "What do you think about telemedicine?"
- **Q2** - "Since you joined the telemonitoring program, what has improved the quality of your life?"

A polarity score ranging in $[-1, 1]$ was inferred by adapting the VADER framework to the Italian language and considered as a numerical feature for each set of answers. In Figure 3, the sentiment polarity with respect to the TDO survey Question Q1 is shown through time. In particular, the polarity intensities for the five different epochs are shown in different colors. The results show an overall positive opinion about telemedicine. In Figure 4, the sentiment polarity with respect to the TDO survey Question Q2 is shown through time. Answers related to this question show a more negative polarity score with respect to Question Q1.

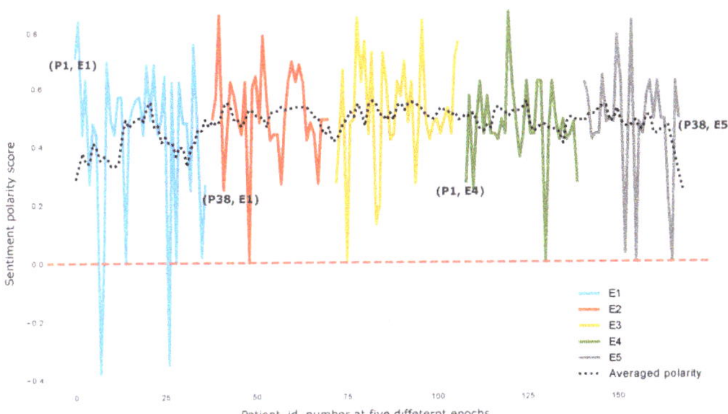

Figure 3. Sentiment polarity associate to Question Q1 are visualized through time. Responses are represented with the relative patient id and the questionnaire session epoch, i.e., E1, E2, E3, etc. On the y-axis, the compound polarity score related to the patient answer at the epoch E_j is inferred by the VADER framework's adaption to the Italian language.

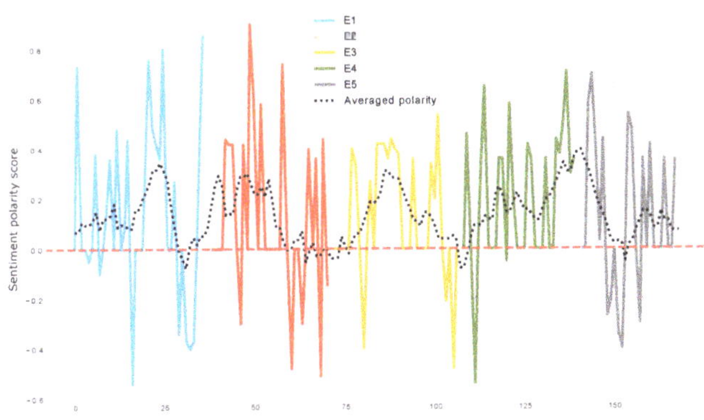

Figure 4. Sentiment polarity associate to Question Q2 are visualized through time. Responses are represented with the relative patient id and the questionnaire session epoch, i.e., E1, E2, E3, etc. On the y-axis, the compound polarity score related to the patient answer at the epoch T_j is inferred by the VADER framework's adaption to the Italian language.

To further provide some insights about the latent aspects more frequently mentioned by patients, in Figure 5, the 50 words resulting more used by patient are shown. The set of free-text answers was pre-processed with standard NLP techniques, i.e., tokenization, stop word removal, and lemmatization.

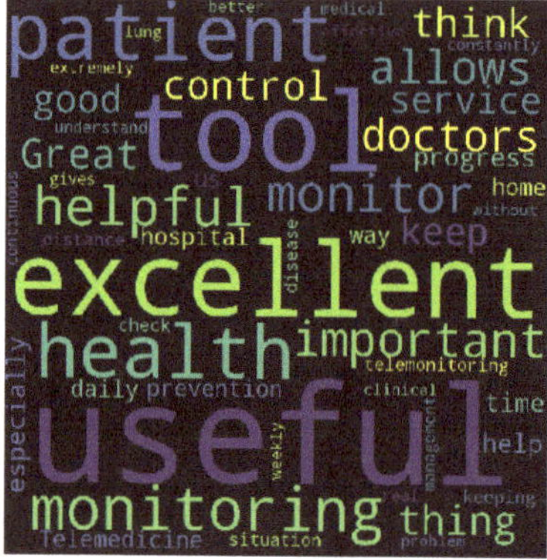

Figure 5. Word cloud showing the most frequent tokens with respect to the answers to question "What do you think about telemedicine?". Tokens with the largest font size are the most frequent.

The results show how "excellent", "useful", "tool", "health", and "patient" are the most common words in the patient response through time.

3.2. Testing Granger-Causality

Three time-series were considered, i.e., polarity scores related to Q1 and Q2 and the time series of Adh-scores. The Augmented Dickey–Fuller Test showed that for all the three considered time series the stationarity condition holds (p-value= 4.6124×10^{-18}, p-value= 3.2185×10^{-7}, and p-value $= 0.0035$, respectively).

Two Granger-causality tests were performed to check whether Q1 Granger-causes Adh-score and whether Q2 Granger-causes Adh-score. Moreover, since all the three series are considered contemporaneously, we also need to check whether Adh-score Granger-causes Q1 and whether Adh-score Granger-causes Q2. Three different test-statistics, i.e., F-test, chi2, and likelihood-ratio, were considered, with the number of lags varying from one to four. Tables 4 and 5 show the results in terms of p-values. It can be seen that both Q1 and Q2 Granger-cause Adh-score for lag = 1. On the other hand, Adh-score appears to not Granger-cause Q1 or Q2.

Therefore, the results suggest the existence of a predictive relationship between the polarity scores series associated with Q1 and the polarity score series associated with Q2 with respect to the Adh-score.

Table 4. Q1 and Adh-score: p-value of Granger-causality test performed with three different statistics and four different lags.

	Q1 $\longrightarrow$ Adh-Score			Adh-Score $\longrightarrow$ Q1		
	F-test	Chi2 test	Likelihood-ratio	F-test	Chi2 test	Likelihood-ratio
1	0.0368	0.0339	0.0350	0.4076	0.4027	0.4031
2	0.0998	0.0908	0.0936	0.6667	0.6585	0.6592
3	0.1063	0.0917	0.0963	0.6628	0.6479	0.6496
4	0.2032	0.1760	0.1833	0.8200	0.8061	0.8064

Table 5. Q2 and Adh-score: *p*-value of Granger-causality test performed with three different statistics and four different lags.

	Q2⟶ Adh-Score			Adh-Score⟶ Q2		
	F-test	Chi2 test	Likelihood-ratio	*F*-test	Chi2 test	Likelihood-ratio
1	0.0020	0.0016	0.0028	0.9970	0.9969	0.9969
2	0.0174	0.0141	0.0155	0.7455	0.7390	0.7394
3	0.0547	0.0446	0.0482	0.8235	0.8148	0.9154
4	0.0865	0.0685	0.0744	0.7566	0.7387	0.7406

3.3. Discussion

The survey instrument and the analysis pipeline were applied to a real case study related to the remote follow-up of patients with cystic fibrosis, held in collaboration with the Cystic Fibrosis Unit, at Children's Hospital "Bambino Gesù", Rome, Italy.

In particular, 169 online surveys sent by 38 patients enrolled in a home telemonitoring program provided by the Cystic Fibrosis Unit at the "Bambino Gesù" Children's Hospital in Rome, Italy, were collected and analyzed through five subsequent questionnaire submissions.

Only answers to two open-ended questions were considered, i.e., Q1 "What do you think about telemedicine?" and Q2 "Since you joined the telemonitoring program, what has improved the quality of your life?".

The time-series of polarity score inferred through the adaption of VADER to the Italian language were used as numerical features to perform the Granger-causality testing model to investigate whether a predictive relationship between the polarity score of open-ended questions and Adh-score may exist.

The experimental results reported in Tables 4 and 5 therefore suggest that, under a Granger-causality perspective, the existence of a predictive relationship between the polarity scores series associated to Q1 and the Adh-score (lag = 1, *p*-value = 0.0339, statistic = Chi2 test) and between the polarity score series associated to Q2 and the Adh-score (lag = 1, *p*-value = 0.0016, statistic = Chi2 test).

The results are consistent with the hypothesis that the polarities extracted from patients' opinion on telemedicine may help predict their average adherence one epoch after the survey administration.

If supported, these results may enable the possibility to intervene early, in a targeted and individual way, to avoid drop-out and continue with the home telemonitoring program which represents a valid aspect of care for patients with Cystic Fibrosis.

Moreover, the early recognition of the reasons that lead the patient to drop-out and intervene immediately may result in:

- better assistance to the cystic fibrosis patient, both clinical and psychological (quality of life and self-management of the disease); and
- improvement of the quality of the telemonitoring program.

4. Conclusions

In the present paper, a system architecture for the extraction of emotional states from textual contents, designed to support the monitoring of patients with chronic disease, is presented. The main goal of the proposed system is to present a methodology to capture the underlying opinions that chronic patients have about the program they are enrolled in and to investigate whether these features may help in the early prediction of patient drop-out from the telemedicine program.

The proposed system is designed in an end-to-end fashion to provide support through the whole process, including the implementation of the questionnaire, the survey administration at scheduled intervals, as well as the analysis. Specific contributions are:

- the design of a self-hosted web-based survey instrument built on top of LimeSurvey for the management of online inquiries over time;
- an analysis pipeline that exploits sentiment analysis techniques to infer a sentiment polarity score for each open-ended answer and uses it as a numerical feature (to the best of our knowledge, this is the first time that this kind of approach has been proposed); and
- the validation of both the survey instrument and the analysis pipeline, which were applied to collect and analyze 169 TDO survey responses sent by 38 patients enrolled in a home telemonitoring program provided by the Cystic Fibrosis Unit at the "Bambino Gesu" Children Hospital in Rome, Italy.

In particular, in the present study, we focused on three variables modeled as time series: the polarity scores extracted from the responses to Question Q1, the polarity scores extracted from the responses to Question Q2, and an adherence score (Adh-score) defined for each epoch starting from the number of spirometries over a three-month window provided by the medical team.

The Granger-causality testing model was performed to assess whether a predictive relationship between the polarity scores series associated to Q1 and the Adh-score (lag = 1, p-value = 0.0339, statistic = Chi2 test) as well as between the polarity score series associated to Q2 and the Adh-score (lag = 1, p-value = 0.0016, statistic = Chi2 test).

Limitations to the present analysis may be found in the small number of data collected up-to-date, which does not allow the investigation of changes for a single patient through time.

Moreover, not every patient answered each survey session, which may have an impact on the number of lags.

Nevertheless, the promising results encourage us to further investigate the potentiality of the proposed architecture and the analysis pipeline with the aim to develop, as future work, a predictive system for the early detection of poorly-adherent patients that may also alert doctors to contact patients and eventually update/personalize their telemedicine program (e.g., in terms of timing, technological equipment, psychological counseling, etc.).

Author Contributions: C.Z. and M.C. conceived the main idea of the algorithm and designed the tests; S.B. and M.C. supervised the design of the system; C.Z. designed the system and ran the experiments; and C.Z. and C.P. collected data; Investigation, S.G. All authors contributed in writing the original draft preparation. All authors have read and agreed to the published version of the manuscript.

Funding: This research received no external funding.

Conflicts of Interest: The authors declare no conflict of interest.

Appendix A

The Telemedicine Drop-Out (TDO) survey

1. What do you think about Telemedicine?
2. What improvement have you had since you have been in Telemonitoring? Choose an answer for each question

I felt safer/more peaceful about my health	Strongly disagree	Disagree	indifferent	Agree	Strongly agree
I felt more followed	Strongly disagree	Disagree	indifferent	Agree	Strongly agree

3. Since you joined the Telemonitoring program, what has improved your quality of life?
4. Since you joined the Telemonitoring program, do you find that the time spent on disease management has decreased?
   □ Yes   □ No
   In yes: Why, choose among the sequel
   □ I do fewer intravenous therapies □ Fewer days of hospitalization
5. Have you ever had difficulty sending spirometry?
   □ Yes, why?   □ No, why?
6. In your opinion, when you don't send spirometry results, what are the related reasons? Choose one or more of the following options
   □ Instrument malfunction   □ Anxiety
   □ Lack of time   □ Forgetfulness
   □ I recently performed spirometry at the   □ I do not feel well
   hospital
   □ Other:
7. When you feel physically unwell, do you find it difficult to transmit?
   □ Yes, why?   □ No, why?
8. When you are emotionally down, do you find it difficult to perform the spirometry test?
9. When you don't send the spirometry test, there are any prevailing emotions related to this event? Choose one or more of the following options
   □ Anxiety   □ Worry
   □ Fear   □ Anger
   □ Sadness   □ Other:
10. If there have been times that you have not performed and / or have not sent spirometry, what prompted you not to do it?
11. Would you like to be contacted after you have transmitted the data?

Strongly disagree	Disagree	indifferent	Agree	Strongly agree

12. Has your relationship with doctors / nurses changed since you followed in Telemonitoring?
   □ Yes, why?   □ No, why?
13. Do you share the sending of data with your family?
   □ Yes, why?   □ No, why?
14. Do you share the sending of data with your family?
   □ Yes, why?   □ No, why?
15. What would you change in the Telemonitoring program you are enrolled in?

References

1. Ryu, S. Telemedicine: Opportunities and Developments in Member States: Report on the Second Global Survey on eHealth 2009 (Global Observatory for eHealth Series, Volume 2). *Healthc. Inf. Res.* **2012**, *18*, 153. [CrossRef]
2. White, L.A.E.; Krousel-Wood, M.A.; Mather, F. Technology meets healthcare: Distance learning and telehealth. *Ochsner J.* **2001**, *3*, 22–29. [PubMed]

3.   Suter, P.; Suter, W.N.; Johnston, D. Theory-based telehealth and patient empowerment. *Popul. Health Manag.* **2011**, *14*, 87–92. [CrossRef] [PubMed]

4.   Nielsen, M.K.; Johannessen, H. Patient empowerment and involvement in telemedicine. *J. Nurs. Educ. Pract.* **2019**, *9*, 54–58. [CrossRef]

5.   Delgoshaei, B.; Mobinizadeh, M.; Mojdekar, R.; Afzal, E.; Arabloo, J.; Mohamadi, E. Telemedicine: A systematic review of economic evaluations. *Med J. Islam. Repub. Iran (MJIRI)* **2017**, *31*, 754–761. [CrossRef]

6.   Armaignac, D.L.; Saxena, A.; Rubens, M.; Valle, C.A.; Williams, L.M.S.; Veledar, E.; Gidel, L.T. Impact of Telemedicine on Mortality, Length of Stay, and Cost Among Patients in Progressive Care Units: Experience From a Large Healthcare System. *Crit. Care Med.* **2018**, *46*, 728. [CrossRef]

7.   Polisena, J.; Tran, K.; Cimon, K.; Hutton, B.; McGill, S.; Palmer, K. Home telehealth for diabetes management: A systematic review and meta-analysis. *Diabetes Obes. Metab.* **2009**, *11*, 913–930. [CrossRef]

8.   Gorst, S.L.; Armitage, C.J.; Brownsell, S.; Hawley, M.S. Home telehealth uptake and continued use among heart failure and chronic obstructive pulmonary disease patients: A systematic review. *Ann. Behav. Med.* **2014**, *48*, 323–336. [CrossRef]

9.   Tagliente, I.; Solvoll, T.; Murgia, F.; Bella, S. Telemonitoring in cystic fibrosis: A 4-year assessment and simulation for the next 6 years. *Interact. J. Med. Res.* **2016**, *5*, e11. [CrossRef]

10.   Sibley, C.D.; Rabin, H.; Surette, M.G. Cystic fibrosis: A polymicrobial infectious disease. *Future Microbiol.* **2006**, *1*, 53–61. [CrossRef]

11.   Bella, S.; Murgia, F.; Tozzi, A.; Cotognini, C.; Lucidi, V. Five years of telemedicine in cystic fibrosis disease. *La Clinica Terapeutica* **2009**, *160*, 457–460. [PubMed]

12.   Crombie, I.K. *Research in Health Care: Design, Conduct and Interpretation of Health Services Research*; John Wiley & Sons: Hoboken, NJ, USA, 1996.

13.   Allery, L.A. Design and use questionnaires for research in medical education. *Educ. Prim. Care* **2016**, *27*, 234–238. [CrossRef]

14.   Popping, R. Analyzing open-ended questions by means of text analysis procedures. *Bull. Sociol. Methodol. De Méthodologie Sociol.* **2015**, *128*, 23–39. [CrossRef]

15.   Schuman, H.; Presser, S. The open and closed question. *Am. Sociol. Rev.* **1979**, *44*, 692–712. [CrossRef]

16.   Schwarz, N. Self-reports: How the questions shape the answers. *Am. Psychol.* **1999**, *54*, 93. [CrossRef]

17.   Friborg, O.; Rosenvinge, J.H. A comparison of open-ended and closed questions in the prediction of mental health. *Qual. Quant.* **2013**, *47*, 1397–1411. [CrossRef]

18.   Zucco, C.; Calabrese, B.; Agapito, G.; Guzzi, P.H.; Cannataro, M. Sentiment analysis for mining texts and social networks data: Methods and tools. *Wiley Interdiscip. Rev. Data Min. Knowl. Discov.* **2019**, e1333. [CrossRef]

19.   Murgia, F.; Cilli, M.; Renzetti, E.; Majo, F.; Soldi, D.; Lucidi, V.; Bella, F.; Bella, S. Remote telematic control in cystic fibrosis. *La Clinica Terapeutica* **2011**, *162*, e121–e124.

20.   Zucco, C.; Bella, S.; Paglia, C.; Tabarini, P.; Cannataro, M. Predicting Abandonment in Telehomecare Programs Using Sentiment Analysis: A System Proposal. In *Proceedings of the IEEE International Conference on Bioinformatics and Biomedicine, BIBM 2018, Madrid, Spain, 3–6 December 2018*; Zheng, H.J., Callejas, Z., Griol, D., Wang, H., Hu, X., Schmidt, H.H.H.W., Baumbach, J., Dickerson, J., Zhang, L., Eds.; IEEE Computer Society: Washington, DC, USA, 2018; pp. 1734–1739. [CrossRef]

21.   Team, L.; Carsten, S. LimeSurvey: An open source survey tool. *LimeSurvey Project* **2012**. Available online: https://www.limesurvey.org/en/ (accessed on 25 November 2020).

22.   McKinney, W. Pandas: A Foundational Python library for Data Analysis and Statistics. *Python High Perform. Sci. Comput.* **2011**, *14*. Available online: https://www.dlr.de/sc/portaldata/15/resources/dokumente/pyhpc2011/submissions/pyhpc2011_submission_9.pdf (accessed on 25 November 2020).

23.   Hutto, C.; Vader, G.E. A parsimonious rule-based model for sentiment analysis of social media text. In Proceedings of the Eighth International AAAI Conference on Weblogs and Social, Ann Arbor, MI, USA, 1–4 June 2014; pp. 1–4.

24.   Dickey, D.A.; Fuller, W.A. Distribution of the estimators for autoregressive time series with a unit root. *J. Am. Stat. Assoc.* **1979**, *74*, 427–431.

25.   Granger, C.W. Testing for causality: A personal viewpoint. *J. Econ. Dyn. Control* **1980**, *2*, 329–352. [CrossRef]

26. Pennebaker, J.W.; Boyd, R.L.; Jordan, K.; Blackburn, K. The Development and Psychometric Properties of LIWC2015. Technical Report. 2015. Available online: https://repositories.lib.utexas.edu/handle/2152/31333 (accessed on 25 November 2020).

27. Bradley, M.M.; Lang, P.J. *Affective Norms for English Words (ANEW): Instruction Manual and Affective Ratings*; Technical Report for C-1; The Center for Research in Psychophysiology, University of Florida: Gainesville, FL, USA, 1999.

28. Basile, V.; Nissim, M. Sentiment analysis on Italian tweets. In *Proceedings of the 4th Workshop on Computational Approaches to Subjectivity, Sentiment and Social Media Analysis, Atlanta, Georgia, 14 June 2013*; Association for Computational Linguistics: Atlanta, Georgia, 2013; pp. 100–107.

29. Pianta, E.; Bentivogli, L.; Girardi, C. MultiWordNet: Developing an Aligned Multilingual Database. In Proceedings of the First International Conference on Global WordNet, Mysore, India, 21–25 January 2002; pp. 293–302.

Publisher's Note: MDPI stays neutral with regard to jurisdictional claims in published maps and institutional affiliations.

MDPI

St. Alban-Anlage 66

4052 Basel

Switzerland

Tel. +41 61 683 77 34

Fax +41 61 302 89 18

www.mdpi.com

Information Editorial Office

E-mail: information@mdpi.com

www.mdpi.com/journal/information